Sputtered Thin Films

Engineering Materials

Series Editor

Kaushik Kumar
Nanomaterials and Nanocomposites: Characterization, Processing, and Applications

B. Sridhar Babu and Kaushik Kumar
Sputtered Thin Films: Theory and Fractal Descriptions

Fredrick M Mwema, Esther T Akinlabi, and Oluseyi P Oladijo
ENGINEERING MATERIALS BOOK SERIES
by CRC Press / Taylor & Francis Publishing Group

Series Editor: Kaushik Kumar

About the Series

This book series focuses on topics of current interest in all aspects of engineering materials. The series aims to be a collection of textbooks, research books, and edited books covering both artificial and natural materials used in engineering and technological applications. It features titles on traditional materials, such as ferrous, nonferrous, ceramic, glass, elastomers, and polymers, as well as advanced materials, including composite materials, smart materials, hierarchical materials, biomaterials, biodegradable materials, functionally graded materials, and shape recognition materials. Topics include macro-, micro-, or nanoscale materials ranging from manufacturing to disposal along with characterization, optimization, applications and many more.

Experts from industry, research institutes, and academia are invited to submit proposals for books, both authored or edited, on all aspects of materials including but not limited to the above.

For inquires or to submit a book proposal, please contact:

Kaushik Kumar, Series Editor (kkumar@bitmesra.ac.in)

Sputtered Thin Films

Theory and Fractal Descriptions

Fredrick Madaraka Mwema,
Esther Titilayo Akinlabi, and
Oluseyi Philip Oladijo

CRC Press
Taylor & Francis Group
Boca Raton London New York

CRC Press is an imprint of the
Taylor & Francis Group, an **informa** business

First edition published 2021
by CRC Press
6000 Broken Sound Parkway NW, Suite 300, Boca Raton, FL 33487-2742

and by CRC Press
2 Park Square, Milton Park, Abingdon, Oxon, OX14 4RN

Library of Congress Cataloging in Publication Data
Names: Mwema, Frederick M. (Frederick Madaraka), author. | Akinlabi, Esther Titilayo, 1976- author. | Oladijo, Oluseyi P. (Oluseyi Philip), author. Title: Sputtered thin films : theory and fractal descriptions / by Frederick M. Mwema, Esther T. Akinlabi, Oluseyi P. Oladijo. Description: First edition. | Boca Raton, FL : CRC Press/Taylor & Francis Group, LLC, 2021. | Series: Engineering materials book series | Includes bibliographical references and index. | Summary: "This book provides an overview of sputtered thin films and demystifies the concept of fractal theory in analysis of sputtered thin films. It simplifies the use of fractal tools in studying the growth and properties of thin films during sputtering processes. Part 1 of the book describes the basics and theory of thin film sputtering and fractals. Part 2 consists of the case studies illustrating specific descriptions of thin films using fractal methods. The book is aimed at engineers and scientists working across a variety of disciplines including materials science and metallurgy as well as mechanical, manufacturing, electrical, and biomedical engineering"-- Provided by publisher.
Identifiers: LCCN 2020049312 (print) | LCCN 2020049313 (ebook) | ISBN 9780367492564 (hbk) | ISBN 9781003053507 (ebk) Subjects: LCSH: Thin films--Mathematical models. | Fractal analysis. | Sputtering (Physics) | Coatings. Classification: LCC TA418.9.T45 M87 2021 (print) | LCC TA418.9.T45 (ebook) | DDC 530.4/175--dc23 LC record available at https://lccn.loc.gov/2020049312LC ebook record available at https://lccn.loc.gov/2020049313

ISBN: 978-0-367-49256-4 (hbk)
ISBN: 978-0-367-51360-3 (pbk)
ISBN: 978-1-003-05350-7 (ebk)

Typeset in Times
by SPi Global, India

For my dad Mwema Muvengei, keep strong
as you continue fighting on that sick bed.
For my wife Joan and twin sons, Robin and Victor.

Fredrick Madaraka Mwema

To my dad of blessed memory - Chief Dare Olorunfemi, sleep on dad!
To my husband - Dr Stephen Akinwale Akinlabi and
my lovely children - Akinkunmi and Stephanie. I love you all.

Esther Titilayo Akinlabi

To the glory of God, the Father, His Son and the Holy Spirit.
and to
My late father - Pa Cosmas Oladijo, continue to rest in peace.

Oluseyi Philip Oladijo

Contents

PART 1   Theory

PART 2   *Typical Studies of Fractal Descriptions of Sputtered Films*

Preface

Thin film technology is key to driving the Industry 4.0 around the world today. The technology finds applications in processing of nanometric materials for various uses. Some of the applications of thin film technology include fabrication of microelectronic devices, sensor devices, biomedical materials and devices, optical devices, machine cutting tools, and so forth. Thin film technology is indeed an advanced manufacturing process and has attracted a lot of interest from the manufacturing sector. There are several techniques of thin film processing, and have been broadly classified into two, chemical vapor deposition (CVD) and physical vapor deposition (PVD) methods. PVD methods are generally preferred due to the possibility of depositing thin films at low temperature conditions. As such, a lot of innovations and studies are currently ongoing, both in industry and academia, to exploit the full potential of the PVD technologies. The main focus of these efforts has been on the parameter–property interactions with the aim of achieving reliable and sustainable thin film processes.

This book contributes, holistically, to these efforts by presenting the applications of fractal theory in sputtering of thin films. In this book, the concept of fractals has been extensively used in thin film characterization to understand their growth mechanisms and to predict the properties of the deposited thin films. The book is presented in two parts and seven chapters.

Chapter 1 presents a detailed theory on thin film growth, structure, and properties. In Chapter 2, the theory of fractals including examples and general applications of fractal in engineering is presented. Methods of fractal measurements in thin films and their detailed mathematical formulations are provided in Chapter 3.

In Chapter 4, typical studies on fractal characterization of hillocks and porosity in sputtered thin films are discussed. In Chapters 5 and 6, mono-fractal and multifractal characterization of sputtered thin films are respectively presented with focus on process parameter–property relationships. Finally, in Chapter 7, fractal theory is presented as a novel tool for predicting the evolution and development of thin film properties.

The book is a very useful resource for researchers and academics in the field of surface engineering and thin film technologies. We hope you shall appreciate it.

Fredrick Madaraka Mwema
University of Johannesburg, South Africa & Dedan Kimathi
University of Technology, Nyeri, Kenya

Esther Titilayo Akinlabi
Pan African University of Life and Earth Sciences (PAULESI), Ibadan, Nigeria

Oluseyi Philip Oladijo
Botswana International University of Science & Technology (BIUST), Palapye,
Botswana & University of Johannesburg, South Africa

MATLAB® is a registered trademark of The Math Works, Inc. For product information, please contact:

The Math Works, Inc.
3 Apple Hill Drive
Natick, MA 01760-2098
Tel: 508-647-7000
Fax: 508-647-7001
E-mail: HYPERLINK "mailto:info@mathworks.com" info@mathworks.com
Web: http://www.mathworks.com

Acknowledgments

The University of Johannesburg through the University Research Committee (URC) and GES 4.0 postdoctoral fellowship is acknowledged for the financial support.

We would also like to acknowledge our families for their support during the preparation of this book. Their understanding and sacrifices have enabled us timely completion of this book.

Authors

Dr. Fredrick Madaraka Mwema is a lecturer and postdoctoral researcher at the Department of Mechanical Engineering at the Dedan Kimathi University of Technology, Kenya, and Department of Mechanical Engineering Science at the University of Johannesburg, South Africa, respectively.

He is currently chair of the Department (COD) of Mechanical Engineering at the Dedan Kimathi University of Technology, Kenya. Dr. Mwema is also the director for the Centre for Nanomaterials and Nanoscience Research at the same university. In terms of teaching, he has over 7 years of undergraduate teaching at university level in Mechanical Engineering and 1 year of postgraduate teaching.

His research interests are the field of manufacturing and materials, specifically in surface engineering and coating technologies, severe plastic deformation, advanced materials, and characterizations. He has extensively worked on thin film preparation through RF magnetron sputtering and published his works in several peer-reviewed journals, book chapters, and conferences. Dr. Mwema has over 60 publications and has written two books. He also has an interest in additive manufacturing technologies including laser-based manufacturing and fused deposition modeling.

Dr. Mwema mentors several postgraduate students in the field of manufacturing and materials. He is currently supervising three masters and two PhD students on additive manufacturing, surface engineering, and advanced recycling of polymer-based composites for green building technologies. He has written several research proposals for funding, some of which have been successful in Kenya.

He is a registered graduate mechanical engineer with Engineering Board of Kenya (EBK) and a member of The American Society of Mechanical Engineers (ASME).

Prof. Esther Titilayo Akinlabi is a full professor of Mechanical Engineering. She assumed office as the director of the Pan African University for Life and Earth Sciences Ibadan (PAULESI), Nigeria in July 2020. Prior to joining PAULESI, she had a decade of meritorious service at the Department of Mechanical Engineering Science, Faculty of Engineering and the Built Environment, University of Johannesburg (UJ), South Africa. During her period of service at UJ, she had the privilege to serve as the head of Department of the Department of Mechanical Engineering Science and as the vice dean for Teaching and Learning for the faculty. Her research interest is in the field of modern and advanced manufacturing processes – Friction stir welding and additive manufacturing. Her research in the field of laser-based additive manufacturing include laser material processing and surface engineering. She also conducts research in the field of renewable energy and biogas production from waste. She is a rated National Research Foundation (NRF) researcher in South Africa and has demonstrated excellence in all fields of endeavors.

Prof. Akinlabi has supervised to completion 22 PhD candidates and 28 masters students. Her leadership, mentorship, and research experience are enviable as she guides her team of postgraduate students through the research journey. She is a recipient of several research grants and has received many awards of recognition to her credit. She is a registered member of the Engineering Council of South Africa (ECSA), Council for the Regulation of Engineering Profession in Nigeria (COREN), South African Institution of Mechanical Engineers (SAIMechE), America Society of Mechanical Engineers (ASME), and the Nigerian Society of Engineers (NSE). Prof. Akinlabi has filed two patents, edited four books, published seven books and authored/co-authored over 500 peer-reviewed publications.

Dr. Oluseyi Philip Oladijo was born in Abeokuta, Nigeria. He received his BEng degree in metallurgical and materials engineering from the Federal University of Technology, Akure, Nigeria. In 2013, he received his PhD degree in Material Science and Engineering field from the University of the Witwatersrand, South Africa. Dr. Oladijo then joined the Department of Chemical, Materials and Metallurgical Engineering at the Botswana International University of Science and Technology (BIUST), as a lecturer, in 2015, and later as a senior lecturer in 2017. He is also a senior research fellow with the University of Johannesburg, South Africa. Currently, his research interests are focused on surface coatings and thin films, tribology, residual stress analysis, advanced characterization techniques using synchrotron and neutron light sources and the manufacturing of fiber composites, from which he has already published several peer-reviewed publications. Dr. Oladijo is a registered member of many professional bodies which include South Africa Institute of Tribology (SAIT), Materials Science and Technology Society of Nigeria (MSN), European Thermal Spraying Association (ETSA), and African Light Source (AFLS).

Part 1

Theory

1 Thin Film Growth, Structure, and Properties

1.1 INTRODUCTION

Thin film is an exciting technology, which finds applications in a variety of fields and industries. Thin films are layers of materials with thicknesses ranging from nanometer to micrometer ranges. The need for nanotechnology in different fields has further enhanced the applications for thin film materials. The technology of thin films dates back to more than 5000 years ago where they were mostly used for decoration purposes. In Egypt for example, the hammering of gold into thin layers of about 0.3 µm has been documented around 1,500 years B.C. [1]. Since then, the science of thin film has significantly advanced in terms of the deposition process and applications. Today, there are many techniques for deposition/production of thin films including vapor deposition methods, magnetron sputtering, atomic layer depositions (ALDs), and chemical methods. These techniques are highly advanced for tuning of the thin films to the desired morphology, properties, and performance. With the advancement in techniques today, it is possible to produce patterned thin films for specific applications [2,3]. Nowadays, thin film technology is applied in the metallization of layers for microelectronics and communication applications, for coating of architectural glass to enhance their optical properties, preparation of thin magnetic films for electronic data storage, and fabrication of thin solar cells, resistors, and dielectrics. Thin films are also being used in energy storage vessels for surface protection against corrosion and as wear-protective films on automotive parts such as spark plug electrodes, turbine blades, pistons, and cylinder surfaces [4].

The growth and formation of thin films are complex processes, and are involving (and complex) to study. With the advancement in imaging technology, researchers in thin film technology have been able to develop models, which are now referred to as structural zone models (SZM), for explaining the growth and formation of thin films [5]. Through the SZMs, the concept of nucleation growth and microstructural evolution of thin films has been studied most especially for physically deposited thin films [5]. One of the most widely accepted models for growth and formation of thin films was developed by Petrov et al. [6] and this model illustrated that the processes (nucleation and grain growth) leading to film formation depend on the underlying process parameters of the deposition technique.

As such, in enhancing the growth and formation of continuous thin films for various applications, researchers are interested in fine-tuning the different parameters affecting the deposition processes. From literature and based on the experience of the authors, there are so many factors influencing the growth and formation of thin films and they can be broadly classified into three, namely,

3

i. the quality and condition of the surface on which the thin films are formed or deposited –substrate,
ii. the quality and condition of the source of the material to be used to form the thin films – target/precursors, and
iii. the condition and quality of the medium/equipment of deposition.

A huge resource of literature is available on the influence of various parameters on the film formation of various materials. For instance, in 2018, the authors of this book published a comprehensive review on the effect of various parameters on the film formation and properties of the pure aluminium thin films prepared through physical methods such as sputtering, ion beam, and thermal spray [5]. It was observed that the deposition of Al thin films through these methods depends on parameters such as vacuum pressure, vacuum gas flow rate, power, temperature, bias voltage, deposition time/rate, and post-deposition annealing [5]. It was determined that tuning the process parameters and deposition conditions enhance the quality, properties, and performance of the Al thin films. The authors have published numerous experimental results in peer-reviewed journals, conferences, and book chapters to demonstrate the influence of parameters of physical deposition methods to pure Al thin films. Some of these articles are cited in this chapter and typical results are included to show the parameters affecting the sputtering technology of thin films [7–10].

1.2 MARKETS AND TRENDS

Thin film materials find application in photovoltaics (PV) solar cells, optical coating, microelectronics board, coatings, and microelectromechanical systems (MEMS), and therefore in understanding the market trends of thin films, demand for the components in such applications should be considered. Due to the increasing adoption and demand for renewable energy (since 2000), International Energy Agency (IEA) has predicted and reported increasing production and installations of thin film PV solar systems [11]. Figure 1.1 shows the world PV production between 2000 and 2011, and as shown, China and Taiwan were the leading producers of thin film PV systems. The production of thin film PV in 2011 was about 30 GW, which is about 10 times the production in 2000.

The data on annual installations of thin films photovoltaic during the same period are shown in Figure 1.2 for the major world markets. As shown, the demand for PV increased considerably between 2000 and 2012, with the European countries accounting for almost 70% of the share of the total annual installations. However, due to a reduction in government subsidies in Europe, there has been a prediction in decline in demand for photovoltaics among the European countries to about 41% in 2020, although the global demand is expected to rise due to the emerging markets in Asia, Africa, and South America [12]. According to IEA 2018 reports, renewable energy will account for more than 70% of the global electricity resource in 2023 with solar photovoltaics accounting for more than 4% [13]. The indoor PV market size, alone, is predicted to rise to 10 US billion dollars from the current 4 US billion dollars [14]. Hence, the demand and market size for PV solar cells is expected to expand

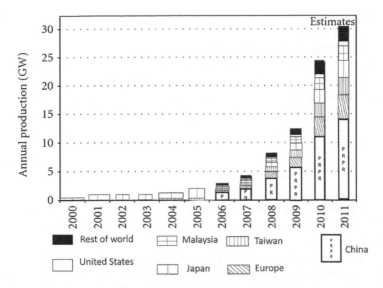

FIGURE 1.1 Data on worldwide photovoltaics production between 2000 and 2011 across the major world producers (*Obtained from Jäger-Waldau, 2012* [11] *under Creative Commons Attribution License*).

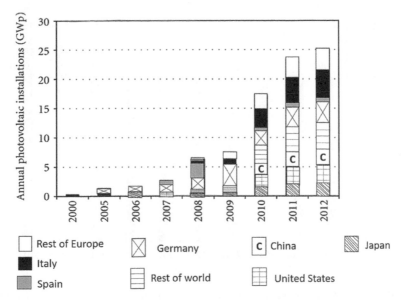

FIGURE 1.2 World data on installations of PV between 2000 and 2012 (*Obtained from Jäger-Waldau, 2012* [11] *under Creative Commons Attribution License*).

exponentially since their average selling price has dropped steadily from US$ 4.12/W in 2008 to US$ 22c/W in 2019 [15].

Internet of things (IoT), fourth industrial revolution (Industry 4.0), and the need for smart systems have pushed the microelectronic industry into a multibillion-dollar sector. For instance, the demand for memory chips increased from 298 to 451 billion US dollars between 2000 and 2018 (https://www.semiconductors.org/). A 10-year billing history for the semiconductor industry in four major world regions is shown in Figure 1.3. It is clearly illustrated that the uptake for semiconductors grew by more than double between 2009 and 2018. The various factors driving the demand for thin films in the electronic industry include (i) the need for miniaturization, low cost, and smart devices, and (ii) requirements for improved ergonomics, artificial intelligence, virtual reality, and wearable devices [16,17]. With the increasing need for smart factories, smart healthcare, self-driving cars, smart TVs, and smart homes, the demand and market for thin films are expected to continue to expand.

The market for thin films is also driven by the demand of sensors for MEMS and wearable devices. For instance, the use of wearable devices in the USA has grown to over 70 million units in 2019. It is predicted that wearable devices will increase from 593 to 929 million units between 2018 and 2021, respectively [18]. The worldwide sales of wearable/smart devices are predicted to increase from 200 million in 2019 to 450 million units in 2022, accounting for about 60% rise in revenue (US billion

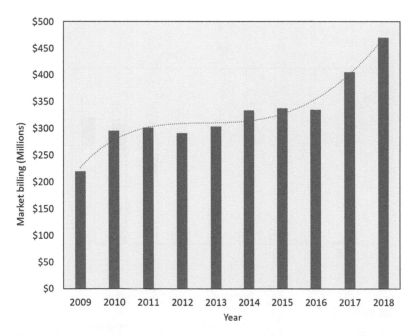

FIGURE 1.3 Historical billing data based on 3-months moving averages for total semiconductor revenues. Data from four semiconductor consumers Americas, Europe, Japan, and Asia-Pacific (*Obtained from https://www.wsts.org/ accessed on 20 August 2019*).

dollars) [18]. According to Wurmser [19], 25% of the US senior citizens (56.7 million people) and 3.8 million young people will utilize a wearable device at least once in every month in 2019. The market for these devices is expected to grow due to the growing number of young people acquiring such devices. For instance, in 2015, more than 24% of youth (ages between 25 and 34) possessed wearable devices with only 6.5% of those between 55 and 64 years old having such devices [19]. The acquisition of these devices is driven by lifestyle (35%), fitness and exercise (29%), health and medical reasons (14%), industrial (10%), entertainment, and gaming (11%) [18,20]. The trends explain the increasing appeal of wearable devices to young people. Some of the most common wearable devices are smartwatches, headbands, wristbands, glasses, and earrings/rings [20].

The market for sensors is predicted to rise between 2019 and 2022 due to (i) the increasing demand for wearable devices, (ii) growing use of IoT, (iii) increasing demand for smartphones, (iv) adoption of industry 4.0, (v) advancements in the sensor technologies, and (vi) the expansion of the automotive industry [19–21]. The increasing adoption of smart systems in modern cars and vehicles coupled with the growing production of vehicles throughout the world implies that there will be more demand for sensors in the automotive industry. The rising need for vehicles with better performance and very high levels of safety means comprehensive monitoring and control of the vehicle systems. The cars of the future will be autonomous, connected, electric, and shared (ACES) and such cars are said to be 'smart'. It is predicted that beyond 2020, 90% of the cars will be connected and shared [22]. Figure 1.4 shows

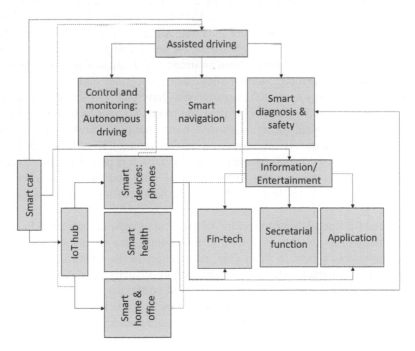

FIGURE 1.4 Features of a smart car and the interrelationships among the various devices (information obtained from Park, Nam, and Kim, 2019 [22]).

the key features of the car of the future (smart). As shown, there are various devices needed to achieve the features of the smart car and all the devices are related to each other in different ways. To achieve a synchronized performance from various devices, several sensors are required, which means that smart cars will considerably contribute to the expansion of the sensor market.

The coating industry is another exciting sector for application of the nanotechnology potential offered by the thin film processes. The introduction of nanoparticles into coating through thin film processes is an attractive alternative to obtain nano-coatings. Such coatings offer a variety of attributes including enhanced composition, excellent chemical stability, higher adhesion strength, better optical properties, smart characteristics (e.g. self-healing and cleaning), compactness and attractive appearance [23,24], and excellent corrosion resistance. Corrosion of metallic substrates is a major issue in the construction and manufacturing industries today and there is an increasing need to developing better coatings and coating materials. In microelectronics, sensors for applications in extreme conditions such as in water, high temperatures, and in the human body should be able to withstand the corrosive behavior of such conditions. As such, the global market for coatings has been on the rise and the market value is expected to rise beyond $191.9 billion by 2023 with a compound annual growth rate (CAGR) of about 5% [25]. The growing market is driven by the increasing consumption from emerging economies in Africa, Asia, and Southern America. Industries such as automotive and aerospace are steadily growing and are likely to expand the global market for coatings and paint. It is expected that by 2024, the global market value of aerospace coating will be over $2496 million [25]. Nano-coating demand will also be driven by the promising opportunities in healthcare, energy, and marine since there is an increasing requirement for superior functionality and performance properties in these sectors. Today, industries need coatings which are self-healing, depolluting, insulative, anticorrosive, anti-fouling, water sheeting, and scratch and UV resistant [26]. Coatings with such characteristics have been fabricated through thin film deposition methods such as chemical and physical vapor techniques. There is also a very high demand for anti-corrosive nano-coatings (thin films) in marine, military, and oil and gas, accounting for a market value of $3 billion in 2012, out of which 1.5% was the nano-coating market. In 2020, it is postulated that the anti-corrosion thin films will account for a CAGR of 9.5% and the total market value will grow to around $6.2 billion out of which the nano-coating market will be $465 million [27].

The need for fractal thin films in the fabrication of solar cells is an indicator of the expansion of the market for thin film technology. In a recent work published in an open-access article in *PLos One*, researchers at the University of Oregon have revealed that fractal-shaped solar panels are visually appealing and reduce stress among the citizens [28]. The study could be the first of a kind combining aesthetics and electrical engineering of solar panels. It has been shown that stress among citizens of the USA costs the economy more than $300 billion per annum. In this study, 370 participants were interviewed on various fractal-like solar panels to determine the most appealing among the population. The study showed that fractal H-shaped solar cells were the most visually appealing as compared to the other shapes. In terms of electrical performance, fractal-shaped solar panels were shown to be superior over non-fractal panels. In 2017, Sandia National Laboratories developed fractal solar power receivers for 1 MW electricity generation [29]. The fractal-like solar power receivers/concentrators were produced through additive manufacturing of Inconel 718 alloy. The manufacturing process was effective for producing and evaluating several fractal designs for solar energy absorption [29]. With these discoveries of novel performance of fractal structures, the market for thin film technology will expand exponentially. This is because most advanced thin film technologies such as sputtering and ALD can produce patterned or fractal-like thin film structures. The authors of the book have demonstrated in several publications that sputtering produces thin films exhibiting fractal characteristics [30,31]. The existence of fractal nature can be demonstrated through analyses of microstructural features such as grain and particle distribution and morphology, and the determination of fractal dimensions for different image scaling or magnification. Additionally, other features such as porosity, hillocks, and other defects in sputtered thin films were shown to exhibit fractal-like behavior in some of the publications by the authors [32] and other researchers [33].

As the preceding discussion illustrates, the demand for thin films is rapidly growing across various sectors. The increasing need for miniaturization, smart systems, and devices, adoption of industry 4.0, and IoT in the modern world implies that the market value for thin films and related technologies will continue to grow and be attractive to multibillion-dollar investments in the next 5 years.

1.3 THEORY OF THIN FILMS AND GROWTH

1.3.1 CLASSIFICATION OF THIN FILM DEPOSITION METHODS

The performance of thin films depends on the method of deposition since each method exhibits different mechanisms and processes to create the thin films. As such, based on the applications and mechanisms, these deposition methods can be classified according to Figure 1.5. As shown, the deposition methods are categorized into two broad categories: (i) those that involve the transformation of the source material into a gaseous state for deposition and (ii) those in which the source material should be in the liquid state for the deposition to take place. In gaseous state methods, the source material (target) is atomized and then transported onto the substrate surface where then it condenses to form a continuous thin film. Usually, enough energy must be created within the source material to cause ionization, melting or atomization, and eventual transport of the source material to the depositing substrate. Different materials will require varied energies for ionization to occur. It is the business of the operator of the thin film deposition equipment to know if the energy is sufficient to ionize and transport the material to the substrate surface. Such determination will depend on the expertise and understanding of the operator with the material properties and the thin film deposition process. Of course, the energy created within the target material is driven by various parameters of the machine, and the combination of those parameters to obtain sufficient energy is very critical. In solution state methods, the source material is usually in liquid form from which it forms thin films through either reaction with the substrate or an electrochemical process. Similarly, the film formation (reactions) is driven by several conditions including the reactants, voltage drop across the electrochemical cell, and condition of the process. The process will also greatly depend on the activation energy of the source materials since the reaction processes, oxidation and reduction, require energies depending on their bonding structures. The conditions of the process, such as temperature, pressure, and conditions of the reactants also play critical roles in driving the thin film deposition processes. In contrast to the first class of methods, these methods may cause environmental pollution since some of the chemicals used may emit fumes, which may irritate the user. However, despite this drawback, chemical-based methods are easier to control especially where patterning and multilayering of films are important.

As shown in Figure 1.5, there are three main types of gaseous/atomistic methods of thin film deposition, namely, chemical vapor deposition (CVD), physical vapor deposition (PVD), and Ion-beam-assisted depositions [34]. In solution state depositions, the three techniques are chemical solution deposition, sol-gel, and electrochemical depositions. As indicated in the figure, various methods give different thicknesses of thin films and therefore the choice of the technique depends on the desired application of the films [35]. To achieve better performance, hybrid deposition processes (in which more than one deposition techniques are utilized for one film) can also be employed. In general, thin film deposition methods involve three significant processes namely (i) synthesis of the source material, (ii) transport of the source material to the desired region of deposition (substrate), and (iii) formation of the thin film [35,36].

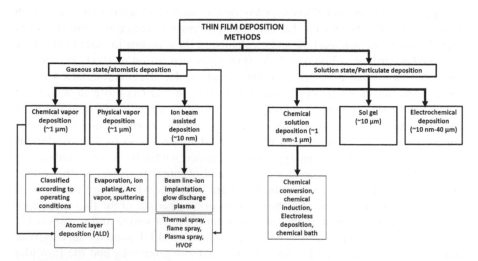

FIGURE 1.5 Classification of thin film deposition techniques (adapted from Abegunde et al., 2019 [35]).

1.3.2 Chemical Vapor Deposition Methods, CVD

CVD involves the deposition of thin films from chemical reactions involving gaseous products. Usually, the gaseous reactants, known as precursors, are kept at high temperatures and are then allowed to react inside a vacuum chamber within the vicinity of the substrate [37]. Figure 1.6 shows the mechanisms of film deposition during CVD [38,39]. The quality and properties of thin films obtained through CVD depend on the experimental conditions, which include the type of the substrate, temperature of reactants and substrate, composition of the precursors, and vacuum pressure [37]. CVD technique is associated with a very high power of deposition of the compounds (reacted precursors) onto the substrate which results in homogeneous thickness and low-porosity thin films. Additionally, CVD can deposit thin films on complex-shaped or patterned substrates since it has the capability for selective deposition (localized reactions). CVDs are classified according to the pressure range and other operating conditions as illustrated in Table 1.1.

CVD has been employed in fabrication of thin films for various applications, including dielectrics [40,41], semiconductors [42,43], tribology and wear protection [44,45], corrosion [46], smart materials [47], and among other applications [37,40].

Among the CVD methods, ALD is one of the most exciting technologies for preparing thin films for a wide range of applications. The method has found applications in the semiconductor processing and it is being used for fabrication of high dielectric constant gate oxides in interconnects [51]. The process allows for atomic level control and manipulation of the thin film growth and formation for specific applications. Unlike the random-based, physical vapor, deposition methods, ALD has been reported to create continuous and porosity-free structures of thin films. The process involves the creation of thin films via reaction of chemical precursors to form a thin film; the reaction may occur on the surface of the substrate or the compounds may be formed off the substrate surface and then deposited on the substrate. The deposition/

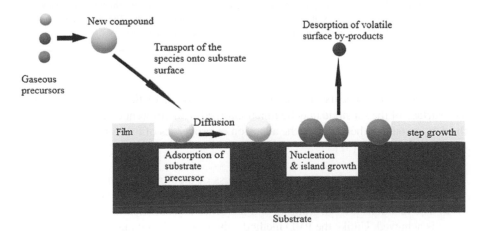

FIGURE 1.6 Processes in a typical chemical vapor deposition of thin films.

TABLE 1.1

Classification of CVD Techniques [37,48–50]

CVD Methods	Description
Atmospheric pressure CVD (APCVD)	It involves high pressure. The CVD processes are undertaken at atmospheric pressure
Low-pressure CVD (LPCVD)	Low/sub-atmospheric pressures
High vacuum CVD (UHVCVD)	Below ~10^{-6} Pa (the processes occur at very low pressures)
Aerosol-assisted CVD (AACVD)	The reactants are conveyed to the substrate through ultra-sonically generated liquid aerosol
Direct liquid injection CVD (DLICVD)	The reactants are in the liquid state, in which they are introduced into the reaction chamber in vapor/gaseous state
Microwave-plasma-assisted CVD (MPCVD)	The process is undertaken in microwave-induced plasma within a chamber at pressures up to 120 torr
Direct-plasma-enhanced CVD (DPECVD)	The process involves the delivery of the precursors into the reaction chamber, generation of the active species for the process through plasma excitation, and CVD activity at the surface of the high temperature substrate
Remote-plasma-assisted CVD (RPECVD)	The process uses plasma to indirectly excite the gas reactants and enhance the thin film deposition. It involves four processes: (i) excitation of the precursors, (ii) extraction of the species necessary for the thin film formation from the plasma, (iii) mixing of the reactants, and (iv) CVD reactions at the surface of the substrate to form thin films. It is a low-temperature process
Atomic layer deposition CVD (ALCVD/ALD)	The process involves the layer-by-layer deposition of materials onto the substrate to create crystalline thin films
Hot wire CVD (HWCVD)	The process involves thermal decomposition of the reactants into radicals using heated tungsten filament. New species, formed from the radicals, get adsorbed onto the surface of the substrate resulting in thin film formation. The substrate is usually kept at low temperatures
Metal–organic CVD (MOCVD) and hybrid physical-CVD (HPCVD)	The CVD methods involving both breakdowns of reactor gases and vaporization of the target solid material
Rapid thermal CVD (RTCVD) and vapor-phase epitaxy (VPE)	The process takes place in a vacuum chamber in which the substrates (wafer) are heated via tungsten–halogen lamps. The equipment and vacuum requirements for these processes are low-cost

creation of the film occurs via the layer-to-layer process of the chemical reactants. The surface of the substrate may be functionalized with different compounds prior to the deposition to either enhance the adhesion or properties of the thin films. ALD has the following merits [51].

i. There is accurate control of the thickness of the films at atomic levels, which ensures that the dimensional accuracy of the deposited films is achieved.
ii. The reactions are self-limiting and therefore nonstatistical deposition process is achieved. Unlike the PVD methods, the process is not random and therefore the surface roughness of the films is lower. Additionally, the self-limiting

nature of the process enhances conformal deposition and excellent step coverage of the substrate surface. There are no sites left during the reaction and therefore continuous films are formed, resulting in very low surface roughness. Thin films with low surface roughness exhibit better electrical and dielectric properties, etc., and better performance.

iii. ALD method is not limited by the size and geometry of the substrates as it may be the case for PVD methods. Since the source materials (precursors) are usually gases, they can fill any space regardless of the nature of the substrate. As such, ALD can create films on substrates with larger surface areas for applications requiring such large coatings and deposition of films on substrates of complex shapes.

1.3.3 Physical Vapor Deposition Method, PVD

PVD involves the transformation of solid materials (target) into a gas state by strafing their surfaces with high-energy ions and transporting their atoms onto the substrate surface. The atoms then condense on the substrate surface to form continuous thin films. The processes occur in vacuum and have been illustrated schematically in Figure 1.7. The main PVD techniques are evaporation, ion plating, arc vapor, and sputtering depositions [35]. Evaporation involves heating a source material (which may be a pure atomic or molecule) to its vaporization temperature inside a vacuum chamber. Usually, the source material (target) is heated to its molten state and then continuous heating causes its vapor to rise above the chamber since the target is usually located below the substrate (or on the lower part of the chamber). The vapor pressure in thermal evaporation is usually maintained between 0.1 and 1 torr although this may depend on the desired rate of evaporation. In high vacuum conditions, the vaporized source materials can easily travel across the chamber and

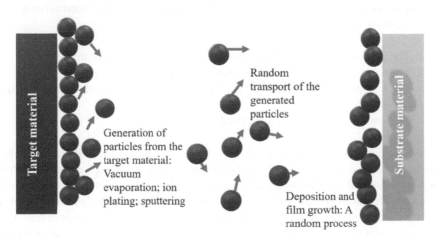

FIGURE 1.7 Mechanisms of PVD techniques. The processes are controlled by temperature, pressure, flowrates of the ionization gases, level of vacuum, etc. [34].

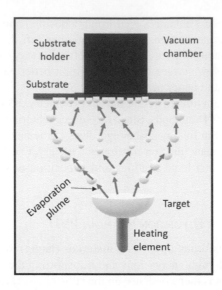

FIGURE 1.8   A schematic diagram of the evaporation of thin film deposition.

deposit onto the substrate with few collisions with any other present particles as shown in Figure 1.8. There are usually two methods of evaporation described in terms of the source of heating of the target material, either using an electric resistive element (thermal evaporation) or electron beam (electron beam evaporation) process. A wide variety of thin films have been prepared through evaporation processes [52–56]. The general focus of most of these reports has been on the influence of deposition parameters on the quality and properties of thermally evaporated films. A few comparative studies between sputtering and thermal evaporation have proven that evaporation is limited by the high level of porosity, defects, and high stress levels obtained on the deposited films [57–59]. The key aspects of thermal evaporation processes are:

i. The melting and vaporization characteristics of the source material,
ii. The temperature of the target and substrate materials,
iii. The vacuum conditions during the thermal evaporation process for thin film deposition, and
iv. The source of power for the thermal spray facility.

Ion plating is an improvement of the evaporation method in which the evaporated materials of the target are ionized by passing them through a plasma before they get deposited onto the substrate (Figure 1.9). This process increases the energy density such that the atoms can stick onto the substrate at higher adhesion than in the normal evaporation process and as such, the process is used to prepare hard thin films for surface protection such as wear resistance, corrosion, and vibration damping properties [60–64]. Similar to other PVD methods, parameters such as the energy of neutrals and ions hitting the surface of the films during deposition

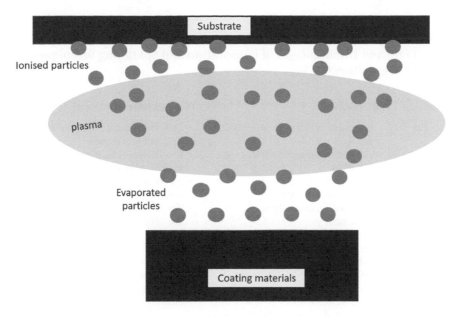

FIGURE 1.9 Ion plating process of thin films.

significantly affect the quality and performance of the thin films [65]. Other important parameters for this process include total pressure inside the deposition chamber, the ion arc current, voltage bias, pulse frequency, partial process pressure, duty ratio, flowrates and pressures of the reactive gases, and axial magnetic field [66]. Due to improved adhesion exhibited by ion plated thin films, the process could be suitable for multilayer film preparations. Such films have attractive applications in extreme conditions such as in surface protection of energy storage vessels, surface cladding of engine parts, and application as coatings in petrochemical pipelines, vessels, etc.

Sputtering is the preferred PVD method for the preparation of thin films due to its simplicity, flexibility, and ability to deposit quality thin films at room temperature. Table 1.2 summarizes the advantages and disadvantages of evaporation and sputtering as the most common PVD methods. Based on the advantages and disadvantages of these common PVD techniques, sputtering stands out as the most preferred since it can prepare thin film from a wide range of materials including ceramics and refractory materials, which are difficult to deposit through evaporation [34]. Additionally, sputtering has been described as an environmentally friendly method and sustainable [67,68]. Compared to evaporation, sputtering has been reported to deposit thin films with less defects such as porosity and cracks and it is not associated with very high residual stresses [57–59]. The process is also associated with high deposition rates and flexibility in terms of parameter and thickness control. These advantages justify the choice of the sputtering method in this book. In the next subsection, details of sputtering technology are described with emphasis on the interrelationships between the parameters and process.

TABLE 1.2

Advantages and Disadvantages of Evaporation and Sputtering Techniques for thin Film Deposition [57,59]

PVD Techniques	Advantages	Disadvantages
Evaporation	• Both metals and non-metals can be deposited • Suitable for low melting point metals and alloys • High rate of depositions • Excellent directionality • Low-cost and can easily be home-made • Easily compatible with plasma for ion-plating process	• Difficult to obtain uniformity in the films • Susceptible to impurity • Low density of the films unless ion-plating is adopted • Not suitable for mass production since it is difficult to scale-up the process • The films produced have higher stresses due to heating involved
Sputtering	• Suitable for metals, non-metals, and dielectrics • Good uniformity of thin films and better step coverage • The level of impurity in thin films is reduced • High density and adhesion films can be obtained • Automation of the process is easier • It is suitable for mass production (high deposition rates for metals) • Less radiation emission	• The deposition rate and hence throughput for dielectrics is low • Poor directionality • High cost of the facility

1.3.4 MECHANISM OF THIN FILM GROWTH

The formation of thin films through the PVD method has been described to occur in four major mechanisms, namely:

i. Nucleation: This process involves the formation of adsorbed monomers or small nuclei of atoms.

ii. Growth of nuclei and crystallization: In this process, the nuclei combine to form small clusters of islands similar to crystallites. The force under which the process occurs is similar to surface tension or capillary action. Usually, the formed clusters are not stable unless a critical temperature is reached where those clusters can achieve a stable size. These clusters are usually interconnected and the structures are characterized by very high porosity.

iii. Coalescence of the clusters to form less connected structures with empty channels: The small islands coalescence to form larger structures and the island density is seen to decrease monotonically at this stage. The process is driven by surface energy, which predominates surface (mass) transfer between adjacent islands, which leads to less interconnections among the structures. The continuous coalescence results into formation networks connected via long, irregular, and narrow empty channels.

iv. Filling of the empty channels: Usually, these channels are exposed regions of the substrate surface. Filling of these channels involves secondary nucleation, growth, and crystallization and coalescence, thereby resulting in a continuous thin film covering the substrate surface.

The evolution of the structure of thin films during the deposition of pure metallic materials has been successfully described via the SZM concept. Generally, most of the SZMs are based on the adatom mobility during the film deposition and growth. According to Petrov et al. [6], the microstructural evolution has three main zones, namely zone I, zone T, and zone II. These zones are differentiated in terms of the adatom mobility and substrate temperature of deposition. The SZM model has been explained by Petrov et al. and others [5,69].

It should be noted that any process, parameter, or condition, which interferes with the above mechanisms of thin film depositions will influence the quality of the grown thin film. As such, the need for understanding, for each of the specific technique, of the relationship between the growth mechanisms and process of deposition cannot be overemphasized.

1.4 THEORY OF SPUTTERING TECHNOLOGY

1.4.1 SCIENCE OF SPUTTERING

Figure 1.10 illustrates the basic theory of sputtering systems. As shown, the process involves the creation of plasma, ejection of material from the target (cathode), transport of the target atoms, deposition, and growth of thin films. The process starts by the formation of ions due to collisions among the neutral atoms and high-energy electrons. Electrons near the cathode are repelled to the anode during which the electrons collide with the neutral gas atoms resulting in more actively (positively)

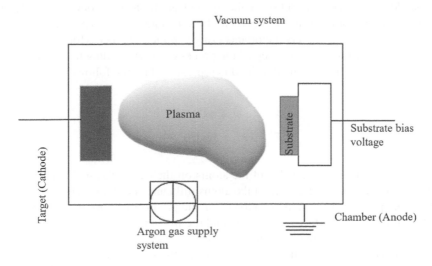

FIGURE 1.10 Basics of the sputtering process.

charged ions. The continuous collision between the electrons and other gas atoms results in further ionization until the gas breaks down. The voltage at which the gas breakdown occurs depends on the pressure in the chamber and anode–cathode separation. When the pressure is too low, there are not enough collisions between the atoms and electrons to sustain the plasma. On the contrary, when the chamber pressure is too high, there will be so many collisions such that the electrons will not compose enough energy to ionise the atoms. Usually, the processes occur in vacuum which is created by a combination of turbo and rotary pumping processes to base pressures of order 10^{-5} Pa and when the inert gas is introduced (usually argon), the vacuum chamber may reach 10^1 Pa. As such, pressure control is critical in a sputtering process for thin film deposition.

The formation of the plasma during the sputtering process is critical since the plasma should possess enough energy to knock the atoms from the surface of the target. The plasma must have at least a threshold energy for the sputtering process to occur. The threshold energy ($E_{threshold}$) for removal of atoms from a target material is defined in Equation (1.1) considering the mass of the plasma atoms (M_1) and mass of the target atoms (M_2).

$$E_{threshold} = \frac{Heat\ of\ vaporization}{\gamma(1-\gamma)}, \quad \text{where}\ \gamma = \frac{4M_1M_2}{(M_1 - M_2)^2} \tag{1.1}$$

If the energy of the incident atoms from the plasma is below the $E_{threshold}$, then sputtering cannot occur. The sputtering is also influenced by the atomic collisions between the ejected atoms and the plasma atoms; the target atoms should possess enough energy to overcome these collisions and reach the substrate surface. The atoms' removal rate from the surface of the target material is measured in terms of the sputtering yield. The sputtering yield (ratio of the ejected target atoms to the incident plasma atoms hitting the target materials) considerably depends on the threshold energy and capability of the ejected atoms to overcome energy losses through atomic collisions as they travel to the substrate surface. This means that the plasma power (energy) is very important during a sputtering process for the generation of atoms and sustenance of the deposition process [70]. According to Sigmund [71], the sputter yield, (Y) near the threshold is represented by the following relationship (Equation 1.2).

$$Y = \frac{3}{4\pi^2}\alpha\frac{4M_1 + 4M_2}{(M_1 + M_2)^2}\frac{E_{threshold}}{U_s} \tag{1.2}$$

The surface binding energy (U_s) of the atoms on the target, threshold energy, mass ratio parameter (∞), and masses of the atoms influence the sputter yield in a typical sputtering process. As seen in Equation (1.2), the plasma ion transfers its momentum to the atoms on the surface of the target and the maximum momentum on the target surface can only be achieved when $M_1 = M_2$. The atom dislodgement can only occur at the target if the momentum transferred from the plasma overcomes the surface binding energy of the atoms on the target. It is therefore important to understand the

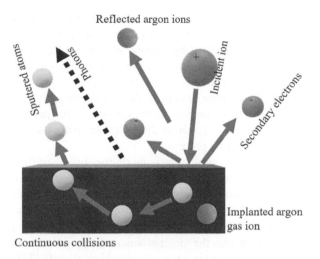

FIGURE 1.11 Various processes taking place at the target during sputtering.

mechanism of formation and sustenance of the plasma during a sputtering process. The plasma formation is described by three regimes, namely dark, glow, and arc discharge regimes, which have been conclusively described in the literature [1] and it will not be discussed in this chapter.

There occurs several mechanisms on the target surface during the sputtering process (Figure 1.11). As shown, when the plasma incident ions hit the target surface, it transfers its kinetic energy onto the target atom which then collides with other atoms continuously leading to collision cascade to eventually dislodge an atom from the surface (sputtering). Additionally, some of the energy is converted into light energy (photon) causing glowing while other plasma ions get reflected away from the target surface or implanted into the target. There may also be the ejection of secondary electrons from the target surface. These electrons can cause further ionization of the argon gas leading to the sustenance of the plasma and hence efficient sputtering. The efficiency of a sputtering process is characterized by the sputtering yield: the number of sputtered target atoms per incident plasma ion. Sputtering yield depends on various factors, some of which include the target type, the binding energy of the target atoms, incident energy of the ion, angle of incident ion, and relative mass of the target atoms [1].

Once the atoms are dislodged from the target, they are supposed to travel to the surface of the substrate for thin film growth and formation where the mechanisms described in the previous subsection (1.3.3) occur. The following are important to note during the transport of the target atoms to the surface of the substrate:

i. The atoms will collide with argon gas atoms inside the vacuum chamber. These collisions change the energies, direction, and momentum of the target atoms.
ii. These changes affect the adatom mobility on the surface of the substrate and as described in Section 1.3.3 (SZM theory), this affects the morphology and microstructure of the grown films.

iii. The mean free path of the sputtered particles depends on the velocity of these particles in relation to the velocity of the argon gas atoms. The theory of mean free path has been described by various authors and extensive literature is available [72]. The free path describes the straight trajectories traveled by the sputtered atoms without collisions by the gas atoms. The presence of many particles within the vacuum chamber reduces the free path for the travel of the target atoms. It, therefore, means that the sputtering free path is also a function of the chamber pressure. The larger the mean free path, the higher the deposition rates of the sputtering process.

iv. In addition to the target particles arriving on the surface of the substrate, other particles resulting from plasma and thermal radiation, compounds (resulting from chemical reactions on the substrate), argon gas atoms, neutral and reflected inert gas atoms, ions, and electrons also impinge the surface of the substrate. The total flux energy on the surface of the substrate is a function of all these particles besides the target atoms and they greatly influence the thin film deposition and growth in a typical sputtering process [71].

1.4.2 SPUTTERING TECHNOLOGIES

The sputtering technology has continuously evolved with time, with various modifications of the process to enhance the sputtering yield and to be able to sputter different types of target materials [4]. The various existing sputtering methods may be classified according to the source of power to the target (cathode) as either direct current (DC) or radiofrequency (RF) sputtering. They may also be classified as either reactive or unreactive sputtering (when reactive gases such as nitrogen or oxygen are used during the sputtering process). If the systems utilize magnetron below the substrate to enhance deposition, it is known as magnetron sputtering (which is further classified as a balanced or unbalanced magnetron sputtering system). Other sputtering technologies include high-power impulse magnetron sputtering (HiPIMS), ion-assisted sputtering, high-target-utilization sputtering (HiTUS), and gas flow sputtering [4, 73]. In the HiPIMS method, large pulsed energy is supplied at the cathode end for a very short time through heavy capacitors. The high power provides a high ionization of the sputtered materials, with a sputtering efficiency of about 90% while ensuring very low heating onto the target [74]. It is the most recent advancement in magnetron sputtering and results in improved properties of the deposited thin films. As such, the method is extensively being used by researchers to deposit high quality thin films including CrN [75], AlTiN [75], and so forth. Ion-assisted sputtering combines the sputtering process with an ion implantation process. In this process, the sputtered atoms are simultaneously and continuously bombarded by a separate flux of ions. In this way, the energy of the sputtered ions is enhanced and hence the quality of the deposited films [76]. The separate source of ions enhances the adhesion of the films onto the substrate and improves the morphology, density, crystallinity, and reduces the residual stresses on the grown films. It has been shown that the energy of the flux can be controlled to tune the size and crystallographic orientation of grains and morphology of the deposited films [76]. There are several recent studies on the deposition of thin films through the ion-assisted sputtering process [77,78]. Each of these processes is used for the deposition of thin films from different target materials and for various applications. Their advantages and disadvantages have been highlighted in the literature [4].

In this book, the presented results were based on the magnetron sputtering method and therefore further details of this process are presented in the next subsection.

1.4.3 MAGNETRON SPUTTERING

Magnetron sputtering is associated with several advantages over other PVD methods, including high deposition rates, high adhesion strength films, and excellent uniformity of the deposited thin films. In this process, magnetic fields are applied onto the cathodic target to confine the plasma within the surface of the target. This ensures there is minimum wastage of the plasma atoms impinging on the target, increases the ionization energy and therefore higher sputtering yield is achieved in magnetron sputtering as compared to the other sputtering technologies [79]. The effect of the magnetrons onto the plasma ensures that sputtering can be undertaken at a much lower voltages (500–600 V) as compared to other sputtering systems which require up to 1 kV.

There are different types of magnetron sputtering systems, which are differentiated by the geometrical variations and positioning of the magnetrons. There are two common configurations of magnetron sputtering systems, that is, conventional and unbalanced magnetrons as shown in Figure 1.12 [80]. In the conventional (balanced) system, the strength of the inner pole balances that of the outer ring of the magnets such that the plasma is strongly confined within the target vicinity. Films prepared through such systems are exposed to concurrent ion bombardment that degrades their structural quality. Furthermore, the substrate is under very low plasma density and

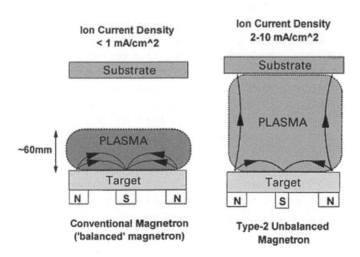

FIGURE 1.12 A schematic illustration of plasma constraining during sputtering in conventional and unbalanced magnetron systems (reproduced with permission from Elsevier copyright number 4927820339245) [80].

the current drawn may not be enough to modify the structure of the coating being deposited on its surface. Although this energy can be increased by increasing the negative bias of the substrate, there is a creation of defects and high stresses on the film. It is, thus, difficult to prepare quality films on large or complex substrates using the conventional magnetron system. In unbalanced magnetrons, the strength of the inner pole is higher than the outer ring (Type-1) or the vice versa (Type-2). In Type-2 unbalanced magnetron (Figure 1.12), the plasma is not fully constrained within the target region but also directed toward the substrate and therefore overcoming the limitations of the conventional sputtering.

1.4.4 Factors Affecting Thin Film Magnetron Sputtering

Magnetron sputtering process is influenced by various processing parameters. Most of these parameters have a direct or indirect relationship with the characteristics of the sputtering plasma inside the deposition chamber. According to literature, these parameters include substrate temperature, DC voltage, DC current, RF power, argon gas flow rate, working pressure, the composition of the target, type of the substrate, deposition time, reactive gases, and among others [5,79]. There are so many publications on the effect of various parameters on magnetron sputtering of different thin films [10,81,82].

Temperature is one of the most significant factors influencing the sputtering process. Most of the sputtering systems have provisions for heating the substrate materials while a few can heat both the substrate and target materials during the deposition process. Increasing the substrate temperature enhances the diffusion of atoms onto the surface of the substrate such that the rate of film growth and deposition increases. High substrate temperature is associated with enhanced crystallinity of the grown thin films. The elevating temperature on the target increases both the energy of the plasma ions impinging on the target and energy of mobility of the target surface atoms. The combination of these effects enhances the sputtering process by increasing the power density on the substrate surface [83]. However, the increased energy of mobility of the particles within the sputtering chamber may cause interference with the mean free path of the process. As such, for each of the different types of material, there is an optimal temperature at which the sputtering process is effective. The use of a hot target during sputtering enhances the deposition rate as it combines the effect of two cathode erosion mechanisms, when the liquid-phase target is used, which include sputtering and evaporation [84]. Sputtering of thick magnetic materials Fe, Ni, Co, etc., is limited since they tend to interfere with the magnetrons of the sputtering system. As such, heating up of such targets above the Curie temperature makes them loose their ferromagnetic properties and behave like paramagnetic materials. That way, they are not able to interfere or shield the magnetic field of the sputtering system, thereby increasing the efficiency of the material removal from the target surface [85].

The types of materials (used for the substrate and targets) are critical determinants of the sputtering process. The threshold energy depends on the type of the target material and type of vacuum gas [86]. The threshold energies of different target materials were reported in an early work by Stuart and Wehner using the

TABLE 1.3
Sputtering Threshold Energies for Selected Materials [87]

Target Material	Incidention Energy (eV) Under Ar ions
Aluminium	13
Titanium	13
Vanadium	23
Chromium	22
Cobalt	25
Nickel	21
Copper	17
Zirconium	22
Silver	15

spectroscopic method under different sputtering ions [87]. Table 1.3 shows the threshold energy of some of the selected materials sputtered under Argon ions as reported by Stuart and Wehner.

Other materials such as hard and magnetic materials of the target also influence the sputtering behavior of the targets [88]. As stated earlier, magnetic materials may influence the magnetic field of the magnetrons leading to difficulty in sputtering. The substrate material also influences the adatom surface mobility and film formation of the target atoms in a sputtering process. Additionally, the substrate type influences the adhesion of the target material onto their surface. For instance, depositing ceramic materials such as AlN on metals, or polymers is associated with lattice mismatch, which causes challenges on the adhesion of the films. The type of substrate also influences the grain evolution of the thin films; for instance, when Al is deposited on metallic substrates, there is formation of well-developed grains, whereas, when Al is deposited on silicon wafers or glass, there is tendency to form porous structures [89].

The vacuum and working pressures are other factors that influence the sputtering process. The vacuum pressure is usually maintained at the order of 10^{-5} torr for the plasma to form. The presence of any gaseous particles inside the chamber may lead to the formation of other products impeding the effective formation of the plasma. Plasma is the 'heart' of the thin film deposition during the sputtering process and therefore sputtering cannot occur if the plasma is not formed and sustained inside the sputtering chamber. As earlier described, the pressure within the deposition chamber during the sputtering process must be maintained at certain levels for the sustainability of the plasma, removal, and transport of the target atoms to the surface of the substrate.

1.5 PARAMETER–PROPERTY RELATIONSHIPS OF SPUTTERED FILMS

As stated in Section 1.4.4., the sputtering process is influenced by various factors associated with the equipment, materials, and environment. The factors associated with the equipment include the operating parameters such as the target holder rotation, target–substrate separation, and power, while those associated with the materials include the chemistry of the materials of the target and substrates and the surface finish of the substrates, whereas those associated with the environment include temperature, gas flow rate, pressure, reactive gases, and so many others. These parameters influence the sputtering process and their influence is measured by the quality of the sputtered thin films. In this regard, the interest of the researchers in the field has been to study the relationship between the sputtering parameters and properties/quality of the produced films.

The influence of the substrate temperature on the sputtered thin films is one of the widely reported aspects of this subject. The authors of this book have extensively researched on the influence of substrate temperature on the deposition of Al thin films on steel substrates through radio-frequency (RF) magnetron sputtering. In one of the studies, titled, 'effect of varying low substrate temperature on sputtered aluminium thin films' published in IOP *Materials Research Express* in 2019, the substrate temperature was varied between 44.5°C and 100°C [10]. It was observed that the structure of the Al thin films prepared at low temperature exhibited amorphous and porous features whereas those deposited at higher substrate temperature exhibited dense, well-defined, and interconnected structures. The atomic force microscopy (AFM) and field emission scanning electron microscopy (FESEM) revealed that the microstructure tends to laterally grow at higher substrate temperature and the columnar structures of Al formed at low temperature shrink to form dense and continuous thin films. It was also reported that the surface microroughness of the Al films decrease with the increase in the substrate temperature. In terms of the mechanical characteristics, the nanoindentation results showed that the hardness and elastic modulus of the Al films are independent of the substrate temperature but rather dependent on the contact depth of the nanoindenter. The increase in substrate temperature from 40°C to 100°C in a relatively different study showed that there is an influence on the preferential crystalline plane orientations [90]. For instance, at 80°C, the (200) crystalline plane was dominant whereas at 100°C, (111) plane was dominant. It was also seen that the (220) and (311) orientations diminished on increasing the substrate temperature. In another study by the current authors, it was reported that the microstructure of the Al thin films deposited on glass substrates evolves with the temperature between 55°C and 95°C [91]. Similar to the films deposited on steel substrates, the films on glass substrates became denser as the temperature of the substrate increased. It was also observed that the nano-hardness and elastic moduli were higher at the maximum substrate temperature of the experimental window. The average vertical surface roughness of the Al thin films on glass substrates was seen to increase with the temperature.

The influence of various process parameters in sputtering method are further detailed in respect to specific studies and materials in Chapters 4–6.

1.6 SPUTTERING OF PATTERNED THIN FILMS

In some applications such as in optics and microelectronics, there is a need to create films exhibiting some special configurations or patterns [92]. In sputtering, patterned thin films can be obtained via two methods as illustrated in Figure 1.13.

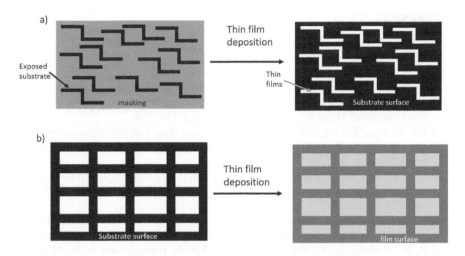

FIGURE 1.13   Routes of sputtering patterned structures of thin films.

   i. Masking: In this method, the surfaces of the substrate which are not supposed to be coated are covered. The cover should have the inverse appearance of the desired surface of the substrate after coating. As shown in Figure 1.13(a), the Z-like profiles represent the sections of the substrates which are not covered but exposed to the coating material. On deposition, the film material (white Z-like profiles) is formed on the exposed substrate structures.

  ii. Creation of patterned porous structures of thin films involves the use of porous structure. On deposition of the thin films, the films have the tendency to nucleate and form on the substrate structures such that the resulting films will exhibit the appearance of the original substrate. This method is illustrated in Figure 1.13(b).

The creation of patterned thin films may result to complex and fractal structures. On the other hand, the theory of fractals can be used to design and study patterned structures of thin films.

1.7 SUMMARY

In this chapter, an overview of the thin film market and trends, methods of thin film fabrication, the theory of sputtering, factors affecting magnetron sputtering, and patterned thin films is presented. As illustrated, the market of thin films continues to expand with the rising demand for smart devices/machines, renewable energy, and the growing adoption of industry 4.0. As such, continuous research into understanding, improvement, and development of thin film fabrication and their properties cannot be overemphasized. The focus of research currently is on the enhancement of the quality of sputtered thin films. As such, studies into the process and performance of thin films are emphasized. The focus of the book is on fractal descriptions of sputtered thin films in relation to performance and deposition conditions. In the next chapters, the theory of fractal, methods of fractal computation, and case studies of fractal descriptions are presented.

REFERENCES

[1] E. Wallin, "Alumina thin films: From computer calculations to cutting tools," Linkoping University, Institute of Technology, 2008.

[2] M. K. Kuntumalla and V. V. S. S. Srikanth, "Surface patterning on nanocrystalline β-SiC thin film by femtosecond laser irradiation," *Mater. Lett.*, vol. 243, pp. 136–139, 2019.

[3] J. S. Lee and W. S. Choi, "Self-patterning methodology by spin coating for oxide thin-film transistors," *Mater. Res. Bull.*, vol. 121, Mar. 2019, 2020.

[4] J. E. Greene, "Review article: Tracing the recorded history of thin-film sputter deposition: From the 1800s to 2017," *J. Vac. Sci. Technol. A Vacuum Surf. Film.*, vol. 35, no. 5, p. 05C204, Sep. 2017.

[5] F. M. Mwema, O. P. Oladijo, S. A. Akinlabi, and E. T. Akinlabi, "Properties of physically deposited thin aluminium film coatings: A review," *J. Alloys Compd.*, vol. 747, pp. 306–323, May 2018.

[6] I. Petrov, P. B. Barna, L. Hultman, and J. E. Greene, "Microstructural evolution during film growth," *J. Vac. Sci. Technol. A Vacuum Surf. Film.*, vol. 21, no. 5, pp. S117–S128, Sep. 2003.

[7] F. M. Mwema, E. T. Akinlabi, O. P. Oladijo, S. Krishna, and J. D. Majumdar, "Microstructure and mechanical properties of sputtered Aluminum thin films," *Procedia Manuf.*, vol. 35, pp. 929–934, 2019.

[8] F. M. Mwema, E. T. Akinlabi, and O. P. Oladijo, "*Influence of sputtering power on surface topography, microstructure and mechanical properties of aluminum thin films,*" in *Proceedings of the Eighth International Conference on Advances in Civil, Structural and Mechanical Engineering – CSM 2019* 2019, Birmingham City, UK, pp. 5–9.

[9] F. M. Mwema, E. T. Akinlabi, and O. P. Oladijo, "Mechanical behaviour of sputtered aluminium thin films under high sliding loads," *Key Eng. Mater.*, vol. 796, pp. 67–73, Mar. 2019.

[10] F. M. Mwema, E. T. Akinlabi, O. P. Oladijo, and J. D. Majumdar, "Effect of varying low substrate temperature on sputtered aluminium films," *Mater. Res. Express*, vol. 6, no. 5, p. 056404, Jan. 2019.

[11] A. Jäger-Waldau, "Thin film photovoltaics: Markets and industry," *Int. J. Photoenergy*, vol. 2012, no. Mar. 2012, p. 2012.

[12] Y. S. Su, "Competing in the global solar photovoltaic industry: The case of Taiwan," *Int. J. Photoenergy*, vol. 2013, no. Jun., 2013.

[13] International Energy Agency, "Renewables, 2018: Market analysis and forecast from 2018 to 2023," 2018.

[14] I. Mathews, S. N. Kantareddy, T. Buonassisi, and I. M. Peters, "Technology and market perspective for indoor photovoltaic cells," *Joule*, vol. 3, no. 6, pp. 1415–1426, 2019.

[15] M. A. Green, "How did solar cells get so cheap?" *Joule*, vol. 3, no. 3, pp. 631–633, 2019.

[16] F. Torrisi and T. Carey, "Graphene, related two-dimensional crystals and hybrid systems for printed and wearable electronics," *Nano Today*, vol. 23, pp. 73–96, 2018.

[17] J. van den Brand et al., "Flexible and stretchable electronics for wearable health devices," *Solid. State. Electron.*, vol. 113, pp. 116–120, Nov. 2015.

[18] J. E. Mück, B. Ünal, H. Butt, and A. K. Yetisen, "Market and patent analyses of wearables in medicine," *Trends Biotechnol.*, vol. 37, no. 6, pp. 563–566, 2019.

[19] Y. Wurmser, "Wearables 2019," *eMarketer*, pp. 1–6, 2019.

[20] M. Dehghani, K. J. Kim, and R. M. Dangelico, "Will smartwatches last? factors contributing to intention to keep using smart wearable technology," *Telemat. Informatics*, vol. 35, no. 2, pp. 480–490, 2018.

[21] S. Y. Yurish, N. V. Kirianaki, and I. L. Myshkin, "World sensors and MEMS markets: Analysis and trends," *Sensors Transducers Mag.*, vol. 62, no. 12, pp. 456–461, 2005.

[22] J. Park, C. Nam, and H. Kim, "Exploring the key services and players in the smart car market," *Telecomm. Policy*, no. April, pp. 1–15, May 2019.

[23] F. Meng and L. Liu, "Electrochemical evaluation technologies of organic coatings," in Jaime Andres Perez Taborda, and Alba Avila (eds.), *Coatings and Thin-Film Technologies*, vol. i, no. tourism, London, UK: IntechOpen, 2019, p. 13.

[24] A. Pruna, "Nanocoatings for protection against steel corrosion," in Fernando Pacheco-Torgal, Maria Vittoria Diamanti, Ali Nazari, Claes Goran Granqvist, Alina Pruna, and Serji Amirkhanian (eds.), *Nanotechnology in Eco-efficient Construction*, Cambridge, UK: Elsevier, 2019, pp. 337–359.

[25] "Global Market Insights releases report on the automotive coating industry," *Focus Powder Coatings*, vol. 2018, no. 12, p. 7, Dec. 2018.

[26] K. Tator, "Nanotechnology: The future of coatings-part 1," *Mater. Perform.*, vol. 53, no. 34–36, pp. 1–4, 2014.

[27] Futuremarketsinc.com, "Construction & exterior protection," *Nanocoatings*, no. April, 2014.

[28] E. T. Roe et al., "Fractal solar panels: Optimizing aesthetic and electrical performances," *PLoS One*, vol. 15, no. 3, p. e0229945, Mar. 2020.

[29] DOE/Sandia National Laboratorie, "New fractal-like concentrating solar power receivers are better at absorbing sunlight," *Sci. Daily*, pp. 8–11, 2017.

[30] F. M. Mwema, E. T. Akinlabi, O. P. Oladijo, and O. P. Oladijo, "The use of power spectrum density for surface characterization of thin films," in Xiao-Yu Yang (ed.), *Photoenergy Thin Film Mater*, Beverly, MA: Scrivener Publishing LLC. pp. 379–411, Mar. 2019.

[31] F. M. Mwema, E. T. Akinlabi, and O. P. Oladijo, "Effect of substrate type on the fractal characteristics of AFM images of sputtered aluminium thin films," *Mater. Sci.*, vol. 26, no. 1, pp. 49–57, Nov. 2019.

[32] F. M. Mwema, E. T. Akinlabi, and O. P. Oladijo, "Fractal analysis of hillocks: A case of RF sputtered aluminum thin films," *Appl. Surf. Sci.*, vol. 489, no. May, pp. 614–623, Sep. 2019.

[33] S. Nazarpour and M. Chaker, "Fractal analysis of Palladium hillocks generated due to oxide formation," *Surf. Coatings Technol.*, vol. 206, no. 11–12, pp. 2991–2997, 2012.

[34] X. Liu, P. K. Chu, and C. Ding, "Surface modification of titanium, titanium alloys, and related materials for biomedical applications," *Mater. Sci. Eng. R Reports*, vol. 47, no. 3–4, pp. 49–121, 2004.

[35] O. O. Abegunde, E. T. Akinlabi, O. P. Oladijo, S. Akinlabi, and A. U. Ude, "Overview of thin film deposition techniques," *AIMS Mater. Sci.*, vol. 6, no. 2, pp. 174–199, 2019.

[36] P. H. Li and P. K. Chu, "Thin film deposition technologies and processing of biomaterials," in Hans J. Griesser (ed.), *Thin Film Coatings for Biomaterials and Biomedical Applications*, Cambridge, UK: Elsevier, 2016, pp. 3–28.

[37] P. O'Brien, "Chemical vapor deposition," in K.H. Jürgen Buschow, Merton C. Flemings, Edward J. Kramer, Patrick Veyssière, Robert W. Cahn, Bernhard Ilschner, and Subhash Mahajan (eds.), *Encyclopedia of Materials: Science and Technology*, Cambridge, UK: Elsevier, 2001, pp. 1173–1176.

[38] J. Nishizawa and H. Nihira, "Mechanisms of chemical vapor deposition of silicon," *J. Cryst. Growth*, vol. 45, no. C, pp. 82–89, Dec. 1978.

[39] L. J. Giling, "Mechanisms of chemical vapour deposition," *Mater. Chem. Phys.*, vol. 9, no. 1–3, pp. 117–138, Sep. 1983.

[40] S. Sivaram, *Chemical Vapor Deposition*, vol. 73. Boston, MA: Springer US, 1995.

[41] J. H. Shin, S. H. Kim, S. S. Kwon, and W. Il Park, "Direct CVD growth of graphene on three-dimensionally-shaped dielectric substrates," *Carbon N. Y.*, vol. 129, pp. 785–789, Apr. 2018.

[42] J. H. Lim, H.-J. Jeong, K.-T. Oh, D.-H. Kim, J. S. Park, and J.-S. Park, "Semiconductor behavior of Li doped ZnSnO thin film grown by mist-CVD and the associated device property," *J. Alloys Compd.*, vol. 762, pp. 881–886, Sep. 2018.

[43] S. Łoś, K. Paprocki, K. Fabisiak, and M. Szybowicz, "The influence of the space charge on The Ohm's law conservation in CVD diamond layers," *Carbon N. Y.*, vol. 143, pp. 413–418, Mar. 2019.

[44] M. Shabani, C. S. Abreu, J. R. Gomes, R. F. Silva, and F. J. Oliveira, "Effect of relative humidity and temperature on the tribology of multilayer micro/nanocrystalline CVD diamond coatings," *Diam. Relat. Mater.*, vol. 73, pp. 190–198, Mar. 2017.

[45] C. Liu, Z. Liu, and B. Wang, "Modification of surface morphology to enhance tribological properties for CVD coated cutting tools through wet micro-blasting post-process," *Ceram. Int.*, vol. 44, no. 3, pp. 3430–3439, Feb. 2018.

[46] R. Sitek, "Influence of the high-temperature aluminizing process on the microstructure and corrosion resistance of the IN 740H nickel superalloy," *Vacuum*, vol. 167, no. November 2017, pp. 554–563, Sep. 2019.

[47] Y. Cui et al., "Thermochromic VO2 for energy-efficient smart windows," *Joule*, vol. 2, no. 9. pp. 1707–1746, Sep. 2018.

[48] G. Lucovsky and D. V. Tsu, "Plasma enhanced chemical vapor deposition: Differences between direct and remote plasma excitation," *J. Vac. Sci. Technol. A Vacuum, Surfaces, Film.*, vol. 5, no. 4, pp. 2231–2238, Jul. 1987.

[49] K. K. Lau, H. Pryce Lewis, S. Limb, M. Kwan, and K. Gleason, "Hot-wire chemical vapor deposition (HWCVD) of fluorocarbon and organosilicon thin films," *Thin Solid Films*, vol. 395, no. 1–2, pp. 288–291, Sep. 2001.

[50] V. E. Borisenko and P. J. Hesketh, *Rapid Thermal Processing of Semiconductors*. Boston, MA: Springer US, 1997.

[51] S. M. George, "Atomic layer deposition: An overview," *Chem. Rev.*, vol. 110, no. 1, pp. 111–131, Jan. 2010.

[52] H. H. Güllü, "Material characterization of thermally evaporated ZnSn2Te4 thin films," *Optik (Stuttg).*, vol. 178, pp. 45–50, Feb. 2019.

[53] H. H. Gullu, M. Isik, and N. M. Gasanly, "Structural and optical properties of thermally evaporated Cu-Ga-S (CGS) thin films," *Phys. B Condens. Matter*, vol. 547, pp. 92–96, Oct. 2018.

[54] E. Aursand and T. Ytrehus, "Comparison of kinetic theory evaporation models for liquid thin-films," *Int. J. Multiph. Flow*, vol. 116, pp. 67–79, Jul. 2019.

[55] C. Sudarshan, S. Jayakumar, K. Vaideki, S. Nandy, and C. Sudakar, "Structural, electrical, optical and thermoelectric properties of e-beam evaporated Bi-rich Bi2Te3 thin films," *Thin Solid Films*, vol. 672, pp. 165–175, Feb. 2019.

[56] S. Lv et al., "Preparation of p-type GaN-doped SnO 2 thin films by e-beam evaporation and their applications in p–n junction," *Appl. Surf. Sci.*, vol. 427, pp. 64–68, Jan. 2018.

[57] G. L. Pakhomov, "Magnetron sputtered vs. thermally evaporated gold contacts in phthalocyanine-based thin film devices," *Microelectronics J.*, vol. 39, no. 12, pp. 1550–1552, Dec. 2008.

[58] H. Garbacz, P. Wieciński, B. Adamczyk-Cieślak, J. Mizera, and K. J. Kurzydłowski, "Studies of aluminium coatings deposited by vacuum evaporation and magnetron sputtering," *J. Microsc.*, vol. 237, no. 3, pp. 475–480, 2010.

[59] R. Koch, "Stress in evaporated and sputtered thin films – a comparison," *Surf. Coatings Technol.*, vol. 204, no. 12–13, pp. 1973–1982, Mar. 2010.

[60] C. X. Tian et al., "Synthesis of monolayer MoNx and nanomultilayer CrN/Mo2N coatings using arc ion plating," *Surf. Coatings Technol.*, vol. 370, pp. 125–129, Jul. 2019.

[61] X. Bai, J. Li, and L. Zhu, "Structure and properties of TiSiN/Cu multilayer coatings deposited on Ti6Al4V prepared by arc ion plating," *Surf. Coatings Technol.*, vol. 372, pp. 16–25, Aug. 2019.

[62] J. Nomoto, T. Tsuchiya, and T. Yamamoto, "Well-defined (0001)-oriented aluminum nitride polycrystalline films on amorphous glass substrates deposited by ion plating with direct-current arc discharge," *Appl. Surf. Sci.*, vol. 478, pp. 998–1003, Jun. 2019.

[63] R. Lan, Z. Ma, C. Wang, G. Lu, Y. Yuan, and C. Shi, "Microstructural and tribological characterization of DLC coating by in-situ duplex plasma nitriding and arc ion plating," *Diam. Relat. Mater.*, vol. 98, p. 107473, Oct. 2019.

[64] D. Guangyu, T. Zhen, B. Dechun, L. Kun, and H. Qingkai, "Vibration damping properties of NiCrAlY coating deposited by arc ion plating," *Rare Met. Mater. Eng.*, vol. 46, no. 5, pp. 1188–1191, May 2017.

[65] M. Maček, M. Čekada, P. Panjan, and M. Mišina, "Effect of reactive gas pressure on the plasma parameters during triode ion plating of Cr–C films," *Czechoslov. J. Phys.*, vol. 50, no. S3, pp. 403–408, Mar. 2000.

[66] C. Peng et al., "Antibacterial TiCu/TiCuN multilayer films with good corrosion resistance deposited by axial magnetic field-enhanced arc ion plating," *ACS Appl. Mater. Interfaces*, vol. 11, no. 1, pp. 125–136, Jan. 2019.

[67] M. Bello and S. Shanmugan, "Achievements in mid and high-temperature selective absorber coatings by physical vapor deposition (PVD) for solar thermal Application-A review," *J. Alloys Compd.*, p. 155510, May 2020.

[68] F. M. Mwema, E. T. Akinlabi, and O. P. Oladijo, "*Sustainability issues in sputtering deposition technology,*" *Proceedings of the International Conference on Industrial Engineering and Operations Management*, Toronto, Canada. pp. 737–744, 2019.

[69] J. E. Yehoda and R. Messier, "Are thin film physical structures fractals?," *Appl. Surf. Sci.*, vol. 22–23, no. PART 2, pp. 590–595, May 1985.

[70] W. Eckstein, C. Garciá-Rosales, J. Roth, and J. László, "Threshold energy for sputtering and its dependence on angle of incidence," *Nucl. Instruments Methods Phys. Res. Sect. B Beam Interact. with Mater. Atoms*, vol. 83, no. 1–2, pp. 95–109, Oct. 1993.

[71] D. Depla, S. Mahieu, and J. E. Greene, "Sputter deposition processes," in *Handbook of Deposition Technologies for Films and Coatings: Science, Applications and Technology*, 3rd ed., P. M. Martin, Ed. Oxford: Elsevier Inc., 2010, pp. 253–295.

[72] W. Posadowski, "Low pressure magnetron sputtering using ionized, sputtered species," *Surf. Coatings Technol.*, vol. 49, no. 1–3, pp. 290–292, 1991.

[73] A. H. Simon, "Sputter processing," in Krishna Seshan, and Dominic Schepis (eds.) *Handbook of Thin Film Deposition*, UK: Elsevier, 2018, pp. 195–230.

[74] V. Jain, T. Roychowdhury, R. G. Kuimelis, and M. R. Linford, "Differences in surface reactivity in two synthetic routes between HiPIMS and DC magnetron sputtered carbon," *Surf. Coatings Technol.*, vol. 378, no. October, p. 125003, 2019.

[75] W. Chen and X. Yan, "Progress in achieving high-performance piezoresistive and capacitive flexible pressure sensors: A review," *J. Mater. Sci. Technol.*, vol. 43, pp. 175–188, Apr. 2020.

[76] R. J. Martín-Palma and A. Lakhtakia, "Vapor-deposition techniques," in *Engineered Biomimicry*, UK: Elsevier, 2013, pp. 383–398.

[77] V. Vishnyakov, P. J. Kelly, J. Humblot, R. J. Kriek, N. S. Allen, and N. Mahdjoub, "Use of ion-assisted sputtering technique for producing photocatalytic titanium dioxide thin films: Influence of thermal treatments on structural and activity properties based on the decomposition of stearic acid," *Polym. Degrad. Stab.*, vol. 157, pp. 1–8, Nov. 2018.

[78] Z. He, S. Zhang, and D. Sun, "Effect of bias on structure mechanical properties and corrosion resistance of TiNx films prepared by ion source assisted magnetron sputtering," *Thin Solid Films*, vol. 676, no. February, pp. 60–67, Apr. 2019.

[79] S. Swann, "Magnetron sputtering," *Phys. Technol.*, vol. 19, no. 2, pp. 67–75, Mar. 1988.

[80] P. J. Kelly and R. D. Arnell, "Magnetron sputtering: A review of recent developments and applications," *Vacuum*, vol. 56, no. 3, pp. 159–172, 2000.

[81] E. Karaköse and H. Çolak, "Effect of substrate temperature on the structural properties of ZnO nanorods," *Energy*, vol. 141, no. 12, pp. 50–55, Dec. 2017.

[82] T. Aubert, M. B. Assouar, O. Legrani, O. Elmazria, C. Tiusan, and S. Robert, "Highly textured growth of AlN films on sapphire by magnetron sputtering for high temperature surface acoustic wave applications," *J. Vac. Sci. Technol. A Vacuum, Surfaces, Film.*, vol. 29, no. 2, p. 021010, Mar. 2011.

[83] E. B. Kashkarov, D. V. Sidelev, M. Rombaeva, M. S. Syrtanov, and G. A. Bleykher, "Chromium coatings deposited by cooled and hot target magnetron sputtering for accident tolerant nuclear fuel claddings," *Surf. Coatings Technol.*, vol. 389, no. February, p. 125618, May 2020.

[84] A. M. Ismailov, V. A. Nikitenko, M. R. Rabadanov, L. L. Emiraslanova, I. S. Aliev, and M. K. Rabadanov, "Sputtering of a hot ceramic target: Experiments with ZnO," *Vacuum*, vol. 168, no. August, p. 108854, Oct. 2019.

[85] V. A. Grudinin et al., "Chromium films deposition by hot target high power pulsed magnetron sputtering: Deposition conditions and film properties," *Surf. Coatings Technol.*, vol. 375, no. June, pp. 352–362, Oct. 2019.

[86] K. Wasa, "Sputtering phenomena," in Kiyotaka Wasa, Isaku Kanno, and Hidetoshi Kotera (ed.), *Handbook of Sputtering Technology*, 2nd ed., UK: Elsevier, 2012, pp. 41–75.

[87] R. V. Stuart and G. K. Wehner, "Sputtering yields at very low bombarding ion energies," *J. Appl. Phys.*, vol. 33, no. 7, pp. 2345–2352, Jul. 1962.

[88] B. Window, "Issues in magnetron sputtering of hard coatings," *Surf. Coatings Technol.*, vol. 81, no. 1, pp. 92–98, May 1996.

[89] F. M. Mwema, E. T. Akinlabi, and O. P. Oladijo, "Microstructure and surface profiling study on the influence of substrate type on sputtered aluminum thin films," *Mater. Today Proc.*, vol. 26, pp. 1496–1499, 2020.

[90] F. M. Mwema, O. P. Oladijo, and E. T. Akinlabi, "Effect of substrate temperature on aluminium thin films prepared by RF-magnetron sputtering," *Mater. Today Proc.*, vol. 5, no. 9, pp. 20464–20473, 2018.

[91] F. M. Mwema, E. T. Akinlabi, and O. P. Oladijo, "Micromorphology of sputtered aluminum thin films: A fractal analysis," *Mater. Today Proc.*, vol. 18, pp. 2430–2439, 2019.

[92] C. Xie et al., "Patterned growth of β-Ga2O3 thin films for solar-blind deep-ultraviolet photodetectors array and optical imaging application," *J. Mater. Sci. Technol.*, vol. 72, pp. 189–196, May 2021.

2 Fractal Theory

2.1 INTRODUCTION

The authors of this book and others have published a powerful review article in a book titled, *Modern Manufacturing Processes* about the applications of fractal theory in modern manufacturing [1]. In the article, the authors described, in detail, the theory of fractals, methods of computing fractal dimensions, and most importantly, different applications of fractal theory in modern manufacturing. The article is the first of a kind on that subject since there are no other reviews presenting applications of fractal in advanced manufacturing processes. Fractal theory and geometry have become accepted in applications in various fields including stock market studies [2], crash rate predictions [3], studying natural systems such as in geosciences [4,5], medicine [6], dentistry [7], surface engineering [8–11], biometrics [12,13], and so forth as illustrated in the review article [1].

The application of the concept of the fractal theory is based on the assumption that processes have self-similar or self-affine characteristics at various scales as exhibited by natural systems such as rivers, mountains, and vegetation cover [1,14]. The concept of self-similarity and self-affinity is a quite confusing subject in fractal geometry, although mathematicians seem to draw a clear distinction between the two terms [15]. It is generally agreed that self-similar objects exhibit the same magnification factor in all the three axes within the same timeframe whereas, self-affine objects are a class of self-similarity objects in which the scaling is different in each of the axis [16,17]. Based on this definition, the review article by Mwema et al. [1] argued that the two terms can be used interchangeably in complex systems, especially in manufacturing, and as such, most of the published articles in manufacturing use the terms concurrently. Throughout this chapter and the entire book, a similar approach is adopted and the terms, self-similar and self-affine, shall be used interchangeably.

As inferred from the literature and experience of the authors of this book on the theory of fractals, the applications of this concept in engineering can be broadly classified into two:

i. To produce fractal-like objects, and
ii. To study/understand the complex formation of structures.

In regards to the first application, there is a tendency in the engineering field to mimic natural systems for better performance of the engineering components. This approach has been known as 'bioinspired human innovation' and it is one of the applications of the patterned thin film concept (Chapter 1). For instance, the honeycomb structure inspired from the bee honeycombs has been used in various applications such in the fields of mechanical engineering, architecture, biomedicine, and

nanotechnology [18,19]. Honeycomb bioinspired devices such as diesel engine fil-
ters and so many have been designed and extensively used in various engineering
applications [20]. Manufacturing technologies such as 3D printing and thin film
deposition have been explored to design and produce bioinspired components such
as those mimicking animals, plants, and rivers. Bioinspired products manufactured
via 3D printing (e.g.. honeycombs) have been studied in literature and have been
shown to exhibit excellent mechanical strength, better thermal, and acoustic characteristics [19].

In regard to the second application, the concept of the fractal theory is used to
understand the growth and scaling behavior of structures manufactured via different processes. For instance, in thin film and surface engineering, the fractal theory
is used in studying the growth mechanism and formation of structures during deposition [21,22]. Through fractal analysis, the growth mechanism (roughening) of
nanostructures during physical vapor deposition of films has been understood; it is
difficult to comprehend the roughness of thin films through the statistical tools
only. It is through fractal techniques such as power spectral density (PSD) functions that the lateral growth (roughening) of the films has been described [23,24].
A book article titled, *The Use of Power Spectrum Density for Surface
Characterization of Thin Film* [11] published by the authors in Wiley, detailed the
step-by-step approach of studying roughness properties of aluminium thin films
deposited via radio-frequency magnetron sputtering. It was discovered that the surface roughness reported via statistical procedures such as root mean square and
average roughness is not fully descriptive of the growth of the films in a typical
physical vapor deposition.

In this chapter, the basics of fractals in relationship to thin film surface analyses
are described. The chapter begins by presenting the definitions and properties of
fractals from the historical and mathematical background, typical examples, in nature
and man-made, of fractals, and general applications of fractals in engineering.
Finally, an overview of the fractal theory of thin film deposition and growth is
presented.

2.2 DEFINITIONS OF FRACTALS AND PROPERTIES

The concept of fractal was developed by Benoit Mendelbrot, a Polish-born French
and an American mathematician, in 1967 while he was trying to determine the length
of the British coastline. However, he was not the one who started the idea of fractals
and as well known in the 19th century, fractal geometry had been used in analytical
and geometric constructions. Mathematical formulations by Joseph Fourier, Karl
Weierstrass, and recently in the 1880's by George Cantor greatly influenced the
development of the concept of fractals in mathematics and other fields. In fact, the
most common Cantor set was coined by Mendelbrot [25]. In his work, Mendelbrot
defined fractal as a fragmented geometry whose smallest units represent the entire
feature of the geometry [26]. With the evolution in its application in different fields,
the definition of fractals has become customized. They are considered infinitely complex structures and have been used to approximate the dimensions of natural systems
such as clouds, coastlines, animal coloration patterns, snowflakes, vegetables such as

broccoli, sea shells, human lungs, and so forth [1]. Generally, fractals have the following characteristics:

i. They are extremely irregular such that the traditional Euclidean geometry cannot be used to describe them. The term fractal originates from the Latin word, fractus, which means irregular surface. Usually, the degree of irregularity is nearly constant at all scales of magnifications.

ii. Fractals consist of fine structures at arbitrarily small scales and as the magnification increases, more details of the structures can be observed.

iii. Fractals have self-similar characteristics, mostly, stochastically and they consist of small replicating building blocks (patterns) equivalent to unit cells in crystallography. As such, the magnified images of fractals are not different from the unmagnified image. This is known as the *magnification invariance* property of fractals.

iv. The fractals have a Hausdorff dimension (fractal dimension) which is usually a non-integer and it is larger than its topological dimension.

v. Finally, fractals have a simple and recursive definition.

Fractal dimension is the most important parameter in fractal analysis in most fields as it represents the complexity of a self-similar object. It is the measure of the number of points which lie on a given set and it is usually a non-integer as it lies between the Euclidean geometry planes [3]. In the book chapter published by the authors on the applications of fractal theory to manufacturing processes [1], the concept of determining of fractal dimension (D) was described and can be written as follows:

$$D = \frac{\log(N)}{\log(r)},$$

where N is the number of self-similar pieces or subdivisions and r is the magnification factor. This dimension is very important in describing natural systems and chaos (trajectories of dynamic systems) and it does not vary with the scale of magnification for fractal structures. Mathematically speaking, the above relationship for computing D indicates that fractal dimension can be calculated by determining the slope of $\log(N)$ against $\log(r)$ and this definition becomes important when dealing with a highly (complex) irregular object, especially for microscopy features.

Historically, fractal dimensions have been determined for some known fractal features. For example, Koch, which consists of a straight line divided into three parts as described earlier [1] has been shown to have a fractal dimension of 1.26. Another fractal feature is the Sierpinski gasket, which consists of equilateral triangles that has been shown to have a fractal dimension of 1.58. A quick comparison between the Koch curve and Sierpinski gasket shows that Sierpinski gasket has a larger value of the fractal dimension. Additionally, as known, the Sierpinski gasket having a complex structure as compared to the Koch curve implies that the complex the object, the larger the fractal dimensions. In other words, a large value of the fractal dimension is an indication of higher roughness (complexity) of the structure. To illustrate this

FIGURE 2.1 Some of the natural fractal objects: (a) Leaf with a fractal dimension of 2.6, (b) forest cover with a fractal dimension of 2.7, (c) broccoli with a fractal dimension of 2.8, and (d) rock with a fractal dimension of 2.75.

argument, we undertook fractal dimensions of images of naturally occurring systems and the results are presented in Figure 2.1.

As shown, the leaf appears to be the simplest and smoothest structure whereas the broccoli appears to be the most complex and rough structure. The observation is physically true for both systems. As an illustration, feeling the surface of the leaf, the rock, and broccoli with one's palm, the leaf feels very smooth, indeed, as compared to the broccoli and rock. Although the broccoli feels smoother than the rock, the fractal dimension indicates that the broccoli is rougher. A close look into the surface of the broccoli compared to the surface of the rock shows that the broccoli surface is more complex than the rock surface. The broccoli surface is composed of very fine details 'bound together' in replication to achieve the over-all structure. A similar argument is true for the vegetation cover as compared to the leaf surface and the other features shown in Figure 2.1. It is can therefore be stated that a larger fractal dimension indicates a complex and rough feature. However, this roughness does not necessarily imply the average height roughness usually computed through the statistical tools. It will be illustrated later in this book that fractal dimension does not always have a direct relationship with the 'height' surface roughness, but the relationship depends on other factors of sur-face structure growth, especially for thin film processes. In more complex sys-tems, especially microstructures of materials, a single fractal dimension value may not be sufficient to represent the roughness of the surfaces. Usually, the fractal dimension for simple systems is constant and a magnified feature repli-cates the characteristics of the whole object. In such cases, the fractal dimension may change due to the physical processes affecting the formation of the surface structures, and such surfaces are said to be multifractal, that is, they exhibit variations in the fractal dimension over their domain. The authors of this book have extensively studied the fractal nature of aluminium thin films and have noted that some of the films exhibited multifractal behaviors. The concept and theory of multifractality will be illustrated later in Chapters 3 and 6 based on the characteristics of thin films.

As the magnification is increased on a fractal feature, the complexity of features increases such that structures that appear 'smooth' at low magnification become jagged. As the magnification is increased further, there are more details obtained and better minute representation of the magnified features for the entire part. In fact, increasing the magnification gives a clearer image of the repeating unit or units contributing to the entire characteristics of the fractal object. Determining the fractal dimension from such features would provide a better approximation. The fractal dimension represents both the lateral and vertical distribution (roughness) of surface features of objects such as the topography of thin film microstructures and natural systems. In a previous book chapter, Mwema et al. [11] demonstrated that for two surfaces with the different lateral distribution of features but with similar height structures, exhibit equal average and root mean square roughness values. It was shown that the use of statistical tools to describe the roughness of surfaces based on height features may be deceiving since they do not consider the lateral roughness/distribution of the surface features. The study showed that PSD was an effective tool in evaluating the surface roughness and distribution of surface features of aluminium thin films deposited on metallic substrates through radio-frequency sputtering.

Because of the potential, the fractal dimension has shown in characterizing objects and processes, several methods have been developed and still so many methods are being considered for accurate approximation of fractal dimensions. Some of the methods of computing fractal dimension based on image analyses include box counting, are-based methods, and Brownian motion methods as they were described by Mwema et al. [1]. The theory and basic mathematical formulations of most of these methods have been provided in Chapter 3. The research focus on this area is currently on enhancing the accuracy of fractal dimension computation, either by enhancing the existing methods or developing advanced methods of determining the fractal dimensions. As such, each of these methods has several mathematical variations based on the applications and the desired accuracy of computing the fractal dimension. As illustrated in the subsequent chapters, the computation of the fractal characteristics of surfaces is mostly based on their images, which may be either be a picture or a magnified feature, and the process involves a few several steps to obtain consistent results.

2.3 EXAMPLES OF FRACTALS

There are several examples of fractal features (either existing in nature or artificial/man made) and they have various applications and uses. Most objects in nature exhibit fractal behavior; and some of the common examples of natural fractals include the clouds, mountains, river connections, biochemical structures, and blood vessels in animals and humans. Trees and crops such as cauliflower, broccoli, ferns, and leaves are more examples of fractals in nature. These natural systems can be modeled via recursive computer algorithms due to the self-similar characteristics. For instance, a branch from a tree is a minute replica of the whole tree although it is not the same but identical. Other natural fractals include snowflake, nautilus shell, gecko's foot, flowerheads of most plants, frost crystals, and lightning bolt (some of these fractals are as shown in Figure 2.2).

The study of the characteristics of natural fractals has proven important in the development of naturally inspired devices for various applications as was stated earlier in the introduction of this chapter. For example, Zhao et al. [27] developed a tree-like photobioreactor through the 3D printing method and illustrated that such design exhibited higher efficiency of bioreaction as compared to conventional designs. Through the nature-inspired fractals, it has been possible to generate artificial fractal structures for better applications and uses in different fields. Some of the artificial fractals include the patterned microstructure of alloys and materials, bird view of cities, tree-like heat exchangers, honeycomb structures, fractal antennae, and microelectronic circuits. Figure 2.3 shows a typical artificial fractal structure of honeycomb design.

FIGURE 2.2 Examples of fractals in nature (a), snowflake (b), unfurling fern (c), fern (d), Nautilus shell (e), gecko's foot (f), Angelic Flowerhead (g), frost crystals (h), lightning bolt (i), river (j), mountains (k), clouds (l), and ocean waves (images were obtained for free from www. pinterest.com).

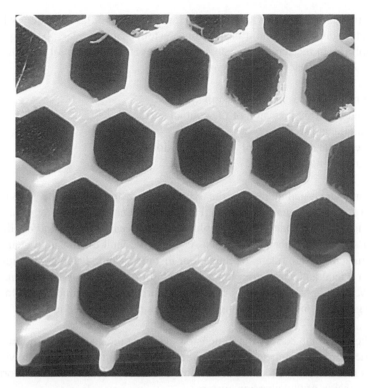

FIGURE 2.3 A typical artificial fractal of honeycomb 3D printed PLA structure for investment mold casting.

2.4 APPLICATIONS OF FRACTAL THEORY IN ENGINEERING

Engineering is a broad field with various disciplines and some of the common fields of engineering include:

    i. Mechanical engineering
    ii. Computer science and engineering
    iii. Electrical and electronic engineering
    iv. Electronics and communication engineering
    v. Civil engineering
    vi. Aeronautical engineering
    vii. Marine engineering
    viii. Chemical engineering
    ix. Biotechnology engineering
    x. Automatic and robotics engineering

In all these fields, there are various applications of fractal theory in understanding some phenomena in related systems or creating naturally inspired devices for specific engineering applications. As reviewed by Han and Qilin [28], fractal theory has been used in studying mechanical transmissions systems such as gears, bearing, pulleys, and couplings. In mechanical transmissions systems, the concept of fractals can find application in the following five areas.

    i. Contact surfaces in transmission systems
    ii. Manufacturing precision of the contact surfaces
    iii. Friction and wear
    iv. Strength and dynamics of the transmission systems
    v. Fault diagnosis

In gear analysis and vibrations, the dynamic characteristics (of the gears) are usually represented by signal waveforms. However, whenever such systems have faults, these characteristics change into complex and nonlinear behavior and exhibit statistics near self-similar vibration signals. As such, the fault diagnosis in gear systems can be studied through fractal methods. For instance, Chun, Deping, and Jiuhua [29] derived a fractal autocorrelation function for fault diagnosis in gear systems. As per the study, and others, the use of fractal theory (autocorrelation function) in fault diagnosis of gear systems, the following important points can be deduced.

    i. The maximum correlation dimension occurs at the no-fault condition.
    ii. The correlation dimension reduces rapidly when a few gear teeth have faults.
    iii. As more and more teeth become faulty, there is a slowdown in the correlation dimension of the vibration system.
    iv. It, therefore, means that the change in correlation dimension in a vibration system indicates a fault in the system and that fractal theory can be used for fault diagnosis in gear systems.

Manufacturing engineering, which is also a discipline in mechanical engineering has extensively employed fractal theory as it was illustrated in the article by the authors of this book [1]. In the article, the author's reviews revealed the following.

    i. In laser manufacturing, the fractal theory is used to understand the hierarchal evolution of laser processed parts at different scales in relationship to processing, performance, and applications. For instance, it was stated that to determine the biomimetic behavior of laser processing, the values of fractal dimensions across different magnifications of parts should be constant. Laser machining quality can also be evaluated through fractal theory as it has been reported in the literature that the higher the values of fractal dimension for such surfaces, the better the surface finish of the laser machined part [30]. Finally, the need for bio-inspired manufacturing has enhanced the applications of additive manufacturing technologies such as laser-based methods. Such technologies are able to produce complex and fractal-like parts.
    ii. In machining processes, fractal theory application has taken a two-fold approach: the topography measurements of machined surfaces [30–32], and signal analyses of cutting forces during the machining process [33]. In the

former, the topographies of machined surfaces are obtained via imaging techniques such as atomic force microscopy, non-contact optical profilometer, and scanning techniques. Then, the microscopy images are taken through image analyses to study their fractal characteristics. In the latter approach, signals obtained during surface machining are direct measurements of the machining process and indicate the cutting forces, wear of the cutting tool, and quality of the machining process. These signals can then be analyzed using fractal techniques such as box counting methods to evaluate their fractal characteristics and understand machining behavior. However, and as observed in the article [1], applications of fractal theory in machining operations in the modern industry is still rare due to the complex nature and unfamiliar chaos theory to the sector.

iii. In friction stir processing, fractal theory application takes a similar approach as the machining methods. Usually, the surface quality of a friction-stir processed surface is obtained through digital imaging, atomic surface microscopy, and surface profilometry, and then the images are analyzed through fractal techniques to understand their fractal behavior. On the other hand, wavelet data obtained are directly related to the machine operating parameters such as force, motor speed, tool vibrations, and so forth. The wavelet data are nonlinear, complex, and exhibit self-similar characteristics, and as such can be studied for fractal behavior.

In robotics, there are five main aspects of applications of fractals, these areas include construction, movement mechanism, computer controls, operating system, and fractal bus. Fractal robots consist of self-similar geometrical parts and require all of the above (five) parts for it to fully operate and in every aspect, the concepts of the fractal theory are used. The construction of such robots involves the design of the smallest unit consisting of all the hardware, electrical, and control systems, and this is followed by an appropriate regeneration of the units into mass for the construction of the entire structure of the robot.

In electrical and electronic fields, the applications of fractal theory have been discussed in various works [34,35], and can be summarized as follows.

i. Fractal theory can be used in the detection of high impedance faults in electrical power systems such as in the sources of power, load, and conductors.

ii. Electronic circuits such as Chua's circuit [34] exhibit chaotic character and nonlinear dynamic behavior. In such circuits, fractal theory can be used to study their characteristics and a lot of literature exists on circuit analyses through fractal methods [36].

iii. Fractal theory is also used in forecasting power loads using fractal dimension and fractal interpolation function theories [37]. Fractal theory is used in load scheduling since load characteristics in a power system is highly dynamic and depends on various parameters like temperature, humidity, and time. The roughness of the load curve can be studied through fractal theory and as such to predict the power loads in power systems.

iv. The theory is also used on studying and analyzing the density of fractal capacitors for various applications in electronic engineering. The density of the fractal capacitor increases as the fractal dimension increases.

v. Lightning is a known natural fractal system as it was shown in Figure 2.2(h). The behavior/properties of lightning are important in various electrical engineering applications. For example, full-scale modeling of an aircraft on the ground requires the use of lightning stroke which is generated via fractal modeling.

vi. In wireless communication, the use of fractal-like antennae has been shown to enhance their performance in terms of transmission and reception of telecommunication signals. These antennae exhibit several benefits for these applications including broadband and multiband frequency response, compact size and shape, mechanical stability, and simplicity in design as compared to the conventional antennae systems.

vii. Sierpinski-shaped impedance transformers have been developed inspired by the Sierpinski fractals. Such transformers have been shown to enhance the operating frequency bandwidth of up to 73% in some literature.

viii. In artificial computer graphics, the fractal theory is used to generate visual effects such as lightning, clouds, rivers, plants, and other living organisms. As such, electronic and computer engineers can utilize these fractal concepts when modeling natural systems for the other disciplines.

In civil engineering, the fractal concept is utilized in simulating natural forms or measuring the complex forms in architectural and structural forms. The availability of powerful computer tools has made it possible for civil engineers to develop highly nonlinear and complex structures for enhanced appeal, mechanical strength, and usability. Through fractal theory, complex building structures such as the hyperbolic paraboloid British Museum (based in London) and Taj Mahal Mausoleum (Figure 2.4)

FIGURE 2.4   Taj Mahal Mausoleum in Uttar Pradesh in India (Courtesy of Tanya Buddi of GRIET, Hyderad, India).

have been developed [38]. Some of the most commonly known fractal structures around the world include the following.

    i. City of Calgary Water Centre
    ii. Milan Cathedral completed in the middle of 19th century
    iii. Beijing National Aquatic Centre
    iv. Staircase in the Vatican museum

In aeronautical engineering, fractal theory finds applications in analyzing the alternative traffic patterns in airspace for flights. It is possible to provide information on the scalability of possible conflicts and the degrees of freedom of the flow of air traffic [39]. Additionally, fractal theory can be used to characterize the complex fracture surfaces of aircraft structures. For instance, Molent et al. [40] demonstrated through fractal analysis of fracture surfaces, it was possible to develop practical models for aircraft structures.

In chemical and process systems engineering, fractals are extensively used to analyze and control chemical processes. This is because chemical processes are characterized by chaos, self-similarity and dynamically nonlinear, and therefore fractal tools can be applied in understanding and controlling these processes [41]. In general, the dynamic behaviors of nonlinear systems are described by a set of algebraic equations, ordinary differential equations, functional or partial equations, or a combination of sets of these equations. Fractal geometry is also used in the analysis of chemical engineering materials. For instance, a recent study by Sarkeheil and Rahbari (2019) [42] titled, 'Fractal geometry analysis of the chemical structure of natural starch modification as a green biopolymeric product' combined box counting technique of fractal method with artificial intelligence for feature extraction and defect detection [42]. The fractal theory has also been used by Ge et al. [43] for the evaluation of porosity and estimation of permeability based on mercury porosimeter data. It was shown that there exists a direct relationship between the structure of the pores and the fractal dimension plots. Additionally, the permeability was shown to have a strong correlation with the fractal parameters. As such, fractal theory can evaluate the characteristics of pores and permeability of chemical structures.

2.5 THIN FILM GROWTH AND FRACTAL THEORY

From the proceeding discussion, the fractal theory has several applications in general engineering and one of the most interesting applications is in thin film technology [44]. In one of the past publications, the author of this book presented a diverse piece of literature on an overview of fractal applications in thin films [9]. Table 2.1 shows the selected publications which were reported in that article and have been reproduced herein with permission from Springer Nature Ltd.

From Table 2.1, the following important notes can be deduced about fractal theory and thin film surface characterizations.

    i. There are several fractal methods applicable for thin film characterization. Some of these include PSD function, height–height correlation function, autocorrelation function, area-based methods, detrended fluctuation analysis

TABLE 2.1

Selected Published Articles and Key Results on the Fractal Analysis of Thin Films (Reproduced with Permission from Springer Nature (Number 4897551022125) from Mwema, Akinlabi, and Oladijo (2019) [9])

Author/Year	Description of the Publication	Fractal Analysis Method(s) Used	Findings and Inferences
Yadav et al. [8]/2014	Fractal characteristics of LiF thin films deposited through electron beam evaporation at 77, 300, and 500 K substrate temperatures were reported	Height–height autocorrelation	The lateral correlation lengths, roughness exponents, and fractal dimensions were computed. The fractal dimension decreased with increasing substrate temperature and roughness. The surfaces of LiF films were shown to be self-affine
Yadav et al. [45]/2017	The study investigated the fractal properties of ZnO prepared by atom beam sputtering on Si at varying deposition angles 20°, 30°, 40°, 60°, 75°	Higuchi's algorithm	The fractal dimension and Hurst exponent were determined. The highest fractal dimensions were reported at 30° and 60° while the lowest at 75°. The highest Hurst exponent was determined at 75°. The surfaces were said to be self-affine
Yadav et al. [46]/2012	The fractal and multifractal analysis were performed on electron beam prepared LiF thin film surfaces at different thicknesses (10, 20, and 40 nm)	Autocorrelation, height–height correlation, and multifractal detrended fluctuation analysis (MFDFA)	The lateral correlation length and fractal dimensions were seen to increase with the film's thickness; whereas the roughness exponent decreased with the thickness of the film
Buchko et al. [47]/2001	The study reports on the fractal analysis of protein polymer films prepared through electrostatic atomization and gas evolution foaming. The films were prepared at different polymer concentrations (1.6–2.4 wt%), the electric field (3–6 kV/cm), deposition separation (1–2 cm), and time (1–10 s).	Power spectral density (PSD)	The fractal dimensions were computed as 2.7 for fiber-only films and 2025 for fiber+bead films. For the bead-only film, the fractal dimension value did not have any physical meaning
Dallaeva et al. [48]/2014	The fractal analysis of AlN epilayers prepared through magnetron sputtering at varying substrate temperatures of 1000, 1300, and 1500 K is reported. The films were deposited on Al_2O_3 substrates	Morphological envelops (cube counting method)	The fractal dimension was computed and shown to increase with the deposition temperature of the substrate

TABLE 2.1 (Continued)

Author/Year	Description of the Publication	Fractal Analysis Method(s) Used	Findings and Inferences
Arman et al. [49]/2015	Fractal analysis of AFM micrographs of Cu films deposited on Si and glass substrates were reported. The films were prepared through DC magnetron sputtering at different thicknesses (5, 25, and 50 nm)	Autocorrelation function, box (cube) counting, and PSD methods	The fractal dimensions were computed and were shown to decrease with the increase in the thickness of the film. The results show that sputtered Cu thin films are self-affine and can be characterized through fractal methods
Talu et al. [50]/2018	The fractal nature of oxidized CdTe surfaces was investigated using AFM	Autocorrelation, height–height and PSD functions	The micromorphology description based on the various fractal techniques was satisfactory to describe the surface oxidation of CdTe
Talu et al. [51]/2016	Surface analysis of Cu/Co nanoparticles prepared by DC magnetron sputtering on Si was undertaken under various power and deposition times	Autocorrelation function	The evolution of correlation lengths, pseudo-topothesy, and fractal dimensions with various deposition parameters were reported
Talu et al. [52]/2016	Fractal analysis of gold nanoparticles in carbon film deposited through RF magnetron sputtering was reported. The sputtering was undertaken under varying power (80–120 W).	Autocorrelation and height–height (structural) functions	Fractal dimensions, pseudo-topothesy, and corner frequency were computed as a function of the sputtering power. There were no proportional relationships among fractal dimension, pseudo-topothesy, and roughness
Yadav et al. [53]/2015	Fractal and multifractal analysis of BaF_2 thin films deposited on Si by electron beam evaporation. The surfaces of the films were irradiated at different ions/cm²	Autocorrelation, multifractal, and PSD functions	Lateral correlation lengths, PSD, roughness exponent, and fractal dimensions were computed
Hosseinpanahi et al. [54]/2015	The paper reported on the fractal and multifractal analysis of CdTe films deposited on glass substrates by sputtering at varying times of 5, 10, and 15 min and a constant power of 30 W	Detrended fluctuation analysis (DFA) and MFDFA	Multifractal parameters and Hurst exponent were computed. The Hurst exponent was shown to be higher than 0.5, indicating a positive correlation and multifractal nature of the films
Nasehnejad et al. [55]/2017	Dynamic scaling analysis of electrodeposited silver thin films is reported. The films were prepared at varying thicknesses (80, 150, 220, 320, 600, and 750 nm)	PSD, height–height correlation	Roughness exponent, power spectrum, and correlation functions were computed. The PSD and correlation functions were seen to increase with the film thickness indicating dynamic scaling of the films. The roughness characteristics were observed to also increase with the film's thickness

(Continued)

TABLE 2.1 (Continued)
Selected Published Articles and Key Results on the Fractal Analysis of Thin Films (Reproduced with Permission from Springer Nature (Number 4897551022125) from Mwema, Akinlabi, and Oladijo (2019) [9])

Author/Year	Description of the Publication	Fractal Analysis Method(s) Used	Findings and Inferences
Davood [56]/2009	The fractal analysis of ITO films prepared through electron beam evaporation was undertaken. The analysis was done on as-deposited, annealed samples (200°C and 300°C)	Box counting method	Fractal dimension was computed for the three AFM micrographs, and higher values were determined on the as-deposited sample. The fractal dimension was shown to decrease with an annealing temperature of the ITO thin films
Davood [57]/2010	A PSD analysis was reported for electron beam evaporated ITO thin films annealed at various temperatures. The annealing temperatures varied as 0°C, 250°C, 350°C, and 450°C	PSD function	The fractal dimensions and slope of the inverse power law were computed. The fractal dimension and roughness were seen to increase with annealing temperature whereas the slope decreased with the annealing temperature
Douketis et al. [58]/1995	The fractal characteristics of vacuum-deposited films at 100 and 300 K were investigated	Cube counting, triangulation, and power spectrum analyses	For the three methods, fractal dimension was computed and averaged. The highest value of fractal dimension was shown to correspond to films with high roughness.
Talu et al. [59]/2018	The study investigated the fractal character of ITO deposited by DC magnetron sputtering under different sputtering chamber conditions (O_2, N_2, and H_2 gases)	Autocorrelation function	Fractal dimension and pseudo-topothesy were computed and shown to vary with different deposition conditions
Mwema et al. [23]/2018	A PSD analysis was undertaken on Al thin films deposited on stainless and mild steel substrates at 150 and 200 W	PSD function	Correlation length, roughness exponent, fractal dimension, and Hurst exponent were determined. The fractal characteristics were seen to evolve with the substrate type and RF power
Li et al. [60]/2000	The fractal model was developed to study the self-affine nature of Co-based thin films	Variation-correlation function	The fractal dimension and correlation length were calculated. The fractal dimension was shown to decrease with an increase in surface roughness. The lowest fractal length was reported at the lowest surface roughness

(DFA) and multifractal detrended fluctuation analysis (MFDFA), and so many others. These methods have been described in Chapter 3 of this book.

ii. The fractal studies on thin films focus on understanding the relationship between thin film growth mechanisms and deposition conditions. Additionally, the studies are also aimed at developing relationships between the vertical roughening and spatial evolution of surface features during thin film deposition.

2.6 SUMMARY

In the chapter, theoretical descriptions regarding the general concept of fractals have been discussed. The definition of fractals and their properties have been provided. Fractals can occur either naturally or artificially and there are various examples in literature. Fractals exhibit self-affinity or/and self-similarity characteristics and therefore their properties are independent of scale. The concept of fractals has been used in general engineering disciplines for mainly two reasons, (i) to mimic natural systems and create bioinspired materials and processes and (ii) to study the behavior of engineering systems. The applications cut across nearly all the engineering fields and therefore fractal theory is such an interesting topic to study. In particular, this book focuses on fractal descriptions of thin films and as illustrated in Section 2.5, the methodology can be used to study the growth mechanisms of thin films during deposition and their interrelationships with deposition parameters. As such, through fractal theory, it is possible to develop models of growth and development of novel thin film structures.

REFERENCES

[1] F. M. Mwema, E. T. Akinlabi, O. P. Oladijo, O. S. Fatoba, S. A. Akinlabi, and S. Tălu, "Advances in manufacturing analysis: Fractal theory in modern manufacturing," in K. Kumar and J. P. Davim (eds.), *Modern Manufacturing Processes*, First., UK: Elsevier, 2020, pp. 13–39.

[2] T. Ikeda, "Multifractal structures for the Russian stock market," *Phys. A Stat. Mech. Its Appl.*, vol. 492, pp. 2123–2128, Feb. 2018.

[3] S. Chand and V. V. Dixit, "Application of Fractal theory for crash rate prediction: Insights from random parameters and latent class tobit models," *Accid. Anal. Prev.*, vol. 112, no. June 2017, pp. 30–38, Mar. 2018.

[4] F. P. Agterberg and Q. Cheng, "Introduction to special issue on 'Fractals and multifractals,'" *Comput. Geosci.*, vol. 25, no. 9, pp. 947–948, Nov. 1999.

[5] B. Klinkenberg, "A review of methods used to determine the fractal dimension of linear features," *Math. Geol.*, vol. 26, no. 1, pp. 23–46, Jan. 1994.

[6] M. N. Starodubtseva, I. E. Starodubtsev, and E. G. Starodubtsev, "Novel fractal characteristic of atomic force microscopy images," *Micron*, vol. 96, pp. 96–102, 2017.

[7] N. B. Nezafat, M. Ghoranneviss, S. M. Elahi, A. Shafiekhani, Z. Ghorannevis, and S. Solaymani, "Topographic characterization of canine teeth using atomic force microscopy images in nano-scale," *Int. Nano Lett.*, vol. 9, no. 4, pp. 311–315, Dec. 2019.

[8] R. P. Yadav, M. Kumar, A. K. Mittal, S. Dwivedi, and A. C. Pandey, "On the scaling law analysis of nanodimensional LiF thin film surfaces," *Mater. Lett.*, vol. 126, no. July, pp. 123–125, 2014.

[9] F. M. Mwema, E. T. Akinlabi, and O. P. Oladijo, "Fractal analysis of thin films surfaces: A brief overview," in Mokhtar Awang, Seyed Sattar Emamian, and Farazila Yusof (ed.), *Advances in Material Sciences and Engineering. Lecture Notes in Mechanical Engineering*, Singapore: Springer, 2020, pp. 251–263.

[10] F. M. Mwema, E. T. Akinlabi, and O. P. Oladijo, "Fractal analysis of hillocks: A case of RF sputtered aluminum thin films," *Appl. Surf. Sci.*, vol. 489, pp. 614–623, Sep. 2019.

[11] F. M. Mwema, O. P. Oladijo, and E. T. Akinlabi, "The use of power spectrum density for surface characterization of thin films," in *Photoenergy and Thin Film Materials*, X.-Y. Yang, Ed. Hoboken, NJ: John Wiley & Sons, Inc., 2019, pp. 379–411.

[12] W.-K. Chen, J.-C. Lee, W.-Y. Han, C.-K. Shih, and K.-C. Chang, "Iris recognition based on bidimensional empirical mode decomposition and fractal dimension," *Inf. Sci. (Ny).*, vol. 221, pp. 439–451, Feb. 2013.

[13] X. Peng, W. Qi, R. Su, and Z. He, "Describing some characters of serine proteinase using fractal analysis," *Chaos, Solitons & Fractals*, vol. 45, no. 7, pp. 1017–1023, Jul. 2012.

[14] R. Lopes and N. Betrouni, "Fractal and multifractal analysis: A review," *Med. Image Anal.*, vol. 13, no. 4, pp. 634–649, Aug. 2009.

[15] T. H. Wilson, "Some distinctions between self-similar and self-affine estimates of fractal dimension with case history," *Math. Geol.*, vol. 32, no. 3, pp. 319–335, 2000.

[16] B. B. Mandelbrot, "Self-Aff ine fractals and fractal dimension," *Phys. Scr.*, vol. 32, pp. 257–260, 1985.

[17] B. B. Mandelbrot, "Self-affine fractal sets, I: The basic fractal dimensions," in Luciano Pietronero and Erio Tosatti (eds.), *Fractals in Physics*, North Holland: Elsevier, 1986, pp. 3–15.

[18] Q. Zhang et al., "Bioinspired engineering of honeycomb structure - Using nature to inspire human innovation," *Prog. Mater. Sci.*, vol. 74. pp. 332–400, Oct. 2015.

[19] J. M. Korde, M. Shaikh, and B. Kandasubramanian, "Bionic prototyping of honeycomb patterned polymer composite and its engineering application," *Polym. Plast. Technol. Eng.*, vol. 57, no. 17, pp. 1828–1844, Nov. 2018.

[20] Z. Han, Z. Jiao, S. Niu, and L. Ren, "Ascendant bioinspired antireflective materials: Opportunities and challenges coexist," *Prog. Mater. Sci.*, vol. 103, pp. 1–68, Jun. 2019.

[21] F. M. Mwema, E. T. Akinlabi, and O. P. Oladijo, "Effect of substrate type on the fractal characteristics of AFM images of sputtered aluminium thin films," *Mater. Sci.*, vol. 26, no. 1, pp. 49–57, Nov. 2019.

[22] S. Stach et al., "3-D surface stereometry studies of sputtered TiN thin films obtained at different substrate temperatures," *J. Mater. Sci. Mater. Electron.*, vol. 28, no. 2, pp. 2113–2122, Jan. 2017.

[23] F. M. Mwema, O. P. Oladijo, T. S. Sathiaraj, and E. T. Akinlabi, "Atomic force microscopy analysis of surface topography of pure thin aluminium films," *Mater. Res. Express*, vol. 5, no. 4, pp. 1–15, Apr. 2018.

[24] K. Ghosh and R. K. Pandey, "Power spectral density-based fractal analysis of annealing effect in low cost solution-processed Al-doped ZnO thin films," *Phys. Scr.*, vol. 94, no. 11, p. 115704, Nov. 2019.

[25] M. L. Lapidus, *Fractal Geometry and Applications—An Introduction to this Volume*, San Diego, CA: American Mathematical Society, vol. 72, no. 1 May 2004.

[26] B. Mandelbrot, "How long Is the coast of Britain? Statistical self-similarity and fractional dimension," *Science (80-.).*, vol. 156, no. 3775, pp. 636–638, May 1967.

[27] L. Zhao et al., "Nature inspired fractal tree-like photobioreactor via 3D printing for CO_2 capture by microaglae," *Chem. Eng. Sci.*, vol. 193, pp. 6–14, 2019.

[28] H. Zhao and Q. Wu, "Application study of fractal theory in mechanical transmission," *Chinese J. Mech. Eng.*, vol. 29, no. 5, pp. 871–879, Sep. 2016.

[29] W. Chun, J. Deping, and W. Jiuhua, "Fractal theory method's research of the gear fault diagnosis," *Adv. Mater. Res.*, 2012, vol. 588–589, pp. 160–165.

[30] H. Chen and Y. Zhou, "Fractal characteristics of 3D surface topography in laser machining," *IOP Conf. Ser. Mater. Sci. Eng.*, vol. 382, no. 4, pp. 1–7. 2018.

[31] J. N. Muguthu and D. Gao, "Profile fractal dimension and dimensional accuracy analysis in machining Metal Matrix Composites (MMCs)," *Mater. Manuf. Process.*, vol. 28, no. 10, pp. 1102–1109, Oct. 2013.

[32] G. Li, K. Zhang, J. Gong, and X. Jin, "Calculation method for fractal characteristics of machining topography surface based on wavelet transform," *Procedia CIRP*, vol. 79, pp. 500–504, 2019.

[33] Z. Jiang, H. Wang, and B. Fei, "Research into the application of fractal geometry in characterising machined surfaces," *Int. J. Mach. Tools Manuf.*, vol. 41, no. 13–14, pp. 2179–2185, Oct. 2001.

[34] A. Jacquin, "An introduction to fractals and their applications in electrical engineering," *J. Franklin Inst.*, vol. 331, no. 6, pp. 659–680, 1994.

[35] A. Tiwari, "Fractal applications in electrical and electronics," *Int. J. Eng. Sci. Adv. Technol.*, vol. 2, no. 3, pp. 406–411, 2012.

[36] L. Lazareck, G. Verch, and A. F. Peter, "*Fractals in circuits*," in *Canadian Conference on Electrical and Computer Engineering 2001. Conference Proceedings (Cat. No.01TH8555)*, Toronto, Ontario, Canada. 2001, vol. 1, pp. 589–594.

[37] L. Jian-Kai, C. Cattani, and S. Wan-Qing, "Power load prediction based on fractal theory," *Adv. Math. Phys.*, vol. 2015, no. 1, pp. 1–6, 2015.

[38] A. Belma and A. Sonay, "*Fractals and fractal design in architecture*," in *13th International Conference "Standardization, Protypes and Quality: A Means of Balkan Countries' Collaboration"*, Brasov, Romania. 2016, pp. 282–291.

[39] S. Mondoloni and D. Liang, "Airspace fractal dimension and applications," 2001.

[40] L. Molent, A. Spagnoli, A. Carpinteri, and R. Jones, "Fractals and the lead crack airframe lifing framework," *Procedia Struct. Integr.*, vol. 2, pp. 3081–3089, 2016.

[41] I. Lee, H. C. Lira, T. Marlin, S. Park, and Y. Yeo, "Full papers applications of chaos and fractals in process," *J. Proc. Cont.*, vol. 6, no. 2, pp. 71–87, 1996.

[42] H. Sarkheil and S. Rahbari, "Fractal geometry analysis of chemical structure of natural starch modification as a green biopolymeric product," *Arab. J. Chem.*, vol. 12, no. 8, pp. 2430–2438, 2019.

[43] X. Ge, Y. Fan, S. Deng, Y. Han, and J. Liu, "An improvement of the fractal theory and its application in pore structure evaluation and permeability estimation," *J. Geophys. Res. Solid Earth*, vol. 121, no. 9, pp. 6333–6345, Sep. 2016.

[44] A. V. Dvornichenko and V. O. Kharchenko, "Scaling properties of the growing monolayer on the disordered substrate," *Phys. Lett. A*, vol. 384, no. 16, p. 126329, Jun. 2020.

[45] R. P. Yadav et al., "Effect of angle of deposition on the Fractal properties of ZnO thin film surface," *Appl. Surf. Sci.*, vol. 416, pp. 51–58, 2017.

[46] R. P. Yadav, S. Dwivedi, A. K. Mittal, M. Kumar, and A. C. Pandey, "Fractal and multifractal analysis of LiF thin film surface," *Appl. Surf. Sci.*, vol. 261, pp. 547–553, 2012.

[47] C. J. Buchko, K. M. Kozloff, and D. C. Martin, "Surface characterization of porous, biocompatible protein polymer thin films," *Biomater. Biomater.*, vol. 22, no. 11, pp. 1289–1300, 2001.

[48] D. Dallaeva, S. Talu, S. Stach, P. Skarvada, P. Tomanek, and L. Grmela, "AFM imaging and fractal analysis of surface roughness of AlN epilayers on sapphire substrates," *Appl. Surf. Sci.*, vol. 312, pp. 81–86, 2014.

[49] A. Arman, Ş. Ţălu, C. Luna, A. Ahmadpourian, M. Naseri, and M. Molamohammadi, "Micromorphology characterization of copper thin films by AFM and fractal analysis," *J. Mater. Sci. Mater. Electron.*, vol. 26, no. 12, pp. 9630–9639, 2015.

[50] Ş. Ţălu, R. Pratap, O. Sik, D. Sobola, and R. Dallaev, "How topographical surface parameters are correlated with CdTe monocrystal surface oxidation," *Mater. Sci. Semicond. Process.*, vol. 85, no. June, pp. 15–23, 2018.

[51] Ş. Ţălu et al., "Microstructure and micromorphology of Cu/Co nanoparticles: Surface texture analysis," *Electron. Mater. Lett.*, vol. 12, no. 5, pp. 580–588, 2016.

[52] Ş. Tălu et al., "Gold nanoparticles embedded in carbon film: Micromorphology analysis," *J. Ind. Eng. Chem.*, vol. 35, pp. 158–166, 2016.

[53] R. P. Yadav, M. Kumar, A. K. Mittal, and A. C. Pandey, "Fractal and multifractal characteristics of swift heavy ion induced self-affine nanostructured BaF2 thin film surfaces," *Chaos An Interdiscip. J. Nonlinear Sci.*, vol. 25, no. 8, p. 083115, Aug. 2015.

[54] F. Hosseinpanahi, D. Raoufi, K. Ranjbarghanei, B. Karimi, R. Babaei, and E. Hasani, "Fractal features of CdTe thin films grown by RF magnetron sputtering," *Appl. Surf. Sci.*, vol. 357, pp. 1843–1848, 2015.

[55] M. Nasehnejad, G. Nabiyouni, and M. G. Shahraki, "Dynamic scaling study of nanostructured silver films," *J. Phys. D. Appl. Phys.*, vol. 50, no. 37, 2017.

[56] D. Raoufi, "Morphological characterization of ITO thin films surfaces," *Appl. Surf. Sci.*, vol. 255, no. 6, pp. 3682–3686, 2009.

[57] D. Raoufi, "Fractal analyses of ITO thin films: A study based on power spectral density," *Phys. B Condens. Matter*, vol. 405, no. 1, pp. 451–455, 2010.

[58] C. Douketis, Z. Wang, T. L. Haslett, and M. Moskovits, "Fractal character of cold-deposited silver films determined by low-temperature scanning tunneling microscopy," *Phys. Rev. B Condens. Matter*, vol. 51, no. 16, pp. 11022–11031, 1995.

[59] S. Ţălu et al., "Micromorphology analysis of sputtered indium tin oxide fabricated with variable ambient combinations," *Mater. Lett.*, vol. 220, pp. 169–171, 2018.

[60] J. M. Li, L. Lu, Y. Su, and M. O. Lai, "Self-affine nature of thin film surface," *Appl. Surf. Sci.*, vol. 161, no. 1, pp. 187–193, 2000.

3   Methods of Fractal Measurements

3.1 INTRODUCTION

There are various methods of computing fractal dimensions for different fields of applications. One of the oldest works by Brian Klinkerberg in 1994 [1] reviewed methods of computing fractal dimensions of one-dimensional curves in the geology field. The author defined some of the one-dimensional objects such as coastlines, river networks, fault traces, and others. Additionally, the author noted that the methods used for linear features could also be used for analyzing more complicated features such as topographic contours of micrographs. There are various methods of computing fractal dimension and studying fractal characteristics of features and some of these methods as enumerated by Klinkerberg include area–perimeter relationship, box counting, divider relationship, Korcak's empirical relationship for islands, line scaling, power scaling, and variogram [1]. In a publication by Nayak, Mishra, and Palai (2019) [2], more methods of fractal analyses were reviewed and it was shown that there are several versions/improvements to the methods described by the Klinkerberg [2]. Besides the above, other methods of fractal analysis include differential box counting (DBC), improved DBC, extended probability box counting, sit island, probabilistic algorithm, covering, and prism methods [2–5].

The fractal methods are defined by different mathematical formulations. However, the accuracy of the computation of fractal characteristics is influenced by related factors as enumerated by Nayak, Mishra, and Palai [2]. These factors include the image properties (quality, pixel range, resolution, texture, self-similarity pattern), sampling parameters (method of sampling, size of steps, number of steps), computation procedure, and mathematical interpretation (curve fitting, regression, etc.). These factors are critical in achieving an accurate estimation of the fractal dimension and other fractal parameters. As a result of these factors, there are continued efforts by the research community on improving the algorithms for different techniques; this has led to inexhaustible methods of studying fractal characteristics of features. In a previous publication, the authors of this book provided an overview of fractal methods for characterizing surfaces of materials in which these methods were classified as box counting, Brownian motion, and area-based methods [6]. In this chapter, methods of analyzing fractal characteristics are described with emphasis on the basic/mathematical formulations and their relevance in characterizing thin film surfaces. The methods discussed in this chapter have been further utilized in analyzing thin film surfaces in the subsequent chapters.

3.2 BOX COUNTING METHOD

The Box counting method is the simplest technique for computing the fractal dimension (D) of fractal features. In this method, the fractal dimension computation is based on the discretization of the space into grids of equal lengths (r). Figure 3.1 shows a typical discretization of an area of a fractal (natural) leaf. Before discretization into equal lengths of grids, the fractal signal should be binarized.

As shown in Figure 3.1, the fractal feature is divided into grids of various sizes and the accuracy of the fractal dimension depends on the size of the grids. As per the box method formulation, the fractal dimension can be determined using Equation (3.1).

$$D = \lim_{r \to 0} \frac{\log N}{\log r} \qquad\qquad (3.1)$$

From the Equation (3.1), the fractal dimension (D) can be determined as the slope of the bi-logarithmic plot of N versus r. The Equation (3.1) represents the simplest box counting formulation, which is applied only in grayscale images; which means that the color images must first be transformed into grayscale before the

FIGURE 3.1 Box counting method for determining the fractal dimension of a natural feature (leaf).

computation. This transformation results in the loss of information from the image. Besides, the box counting method is associated with the following additional limitations.

i. The method ignores the correlation among the intercepted data points on each scale, which are usually correlated positively. In a regression analysis, it is assumed that the data are independent and ignoring this fact in box counting causes underestimation of the confidence interval of the computed fractal dimensions.

ii. It was stated earlier in this section that the accuracy of approximation depends on the number (N) of grids. However, very large N can result into an unrealistically small value of the fractal dimension. Furthermore, the largest grid (N=1) containing all sets of grids gives an inaccurate contribution to the values of the fractal dimensions. It should be noted that for accurate estimation of the fractal dimension, there should be a good scaling in the whole set and a good linear fit of the bi-logarithmic plot of the size and number of grids should be obtained [7,8].

iii. The method does not have a well-defined procedure to evaluate the statistical self-similarity of the objects under investigation. It is usually assumed that the object exhibits a fractal character before it is subjected to the box counting method.

iv. As mentioned, the object is covered by various boxes, and the size and number of boxes intercepting the object under investigation. It was stated earlier that the minimum number of boxes necessary to cover the object should be used. However, a quantification error may arise if there is a miscounting of the minimum number of boxes. Additionally, the location and orientation of these grids are arbitrary and may also lead to the quantification error [9].

To overcome some of the above limitations of the box counting method, there have been various modifications into the method. One such method is the DBC, which has been extensively described in the literature and its formulation is summarized as follows [10]. The method uses grayscale images without binarization and maps the images into a three-dimensional plane. In other words, the grayscale image (I) is divided into a series of cubes. The size of the image is A × A pixels and it is discretized into grids of sizes k × k pixels. If k is assumed to be an integer and ranges between 2 and $\frac{A}{2}$, then the scale, r, can be computed as $r = \frac{k}{A}$. In cases where the k is not divisible by A, then the pixels at the boundary grids are treated as zero. If each grid is covered by a certain number of cubes (n_r) of sizes k × k × c (where c is the height of the cube and denotes the intensity of the grayscale), then the number of cubes (n_r) can be determined using the Equation (3.2).

$$n_r\left(i,j\right) = \left[\frac{g_{max}}{c}\right] - \left[\frac{g_{min}}{c}\right] + 1 \qquad (3.2)$$

In Equation (3.2), g_{max} and g_{min} denotes the maximum and minimum gray-levels on the (i,j) the box, respectively. The height of the grayscale c can be determined as shown in Equation (3.3).

$$c = \frac{k \times G}{A} \tag{3.3}$$

Where G is the total number of grayscale levels in the $I_{A \times A}$ image. It should be noted that Equation (3.2) describes the number of boxes covering one grid on the grayscale image. To determine the total number of boxes (N) to cover the entire image (or all the boxes on the partitioned image), then a summation of all the n_r within the range of scale (r) should be applied as shown in Equation (3.4).

$$N = \sum_{i,j} n_r (i,j); \frac{2}{A} \leq r \leq \frac{1}{2} \tag{3.4}$$

The fractal dimension is determined by plotting a log-log scale of N against 1/r and computing the slope by fitting a straight line using the method of least squares [11]. Suppose there are X points (xi, yi) for the plot and the line of fit is given as $y = Mx + C$, where M is the slope of the line and C is the y-intercept. The method of least squares minimizes the squared errors as shown in Equation (3.5).

$$E = \sum_{i=1}^{X} (yi - Mxi - C)^2 \tag{3.5}$$

Mathematically, to minimize E, the differentials with respect to M and C are undertaken and equated to zero as follows:

$$\frac{dE}{dM} = 0; \text{ and } \frac{dE}{dC} = 0 \tag{3.6}$$

Solving Equation (3.6), yields the following expressions for M and C:

$$M = \frac{X \sum xiyi - \sum xiyi}{X \sum xi^2 - (\sum xi)^2} \tag{3.7}$$

$$C = \frac{\sum xi^2 \sum yi - \sum xiyi \sum xi}{X \sum xi^2 - (\sum xi)^2} \tag{3.8}$$

Therefore, the error in this case can be determined using the expression in Equation (3.9) [10].

$$DE = \frac{1}{X} \sqrt{\frac{\sum_{i=1}^{X} (Mxi + C - yi)^2}{1 + M^2}}; \log(N) = M\log\left(\frac{1}{r}\right) + C \tag{3.9}$$

In Equation (3.9), X is the total number of points used in the curve fitting and M and C are gradient and y-intercept, respectively. As can be seen in this formulation, the parameter c, or the cube height is an important factor influencing the accuracy of the

DBC. A very large value of c gives a wrong estimate of the fractal dimension whereas a very small value of c gives an accurate estimate of the dimension. However, very small values imply a larger computation memory and longer times. As such, a study by Panigraphy et al. [11] investigated the influence of box height (parameter c) on the accuracy of the DBC in which a new method of determining the box height was formulated based on the Brodatz database for Brownian motion; the method was shown to give better results as compared to the existing methods.

The DBC method has been reported to exhibit the following limitations.

  i. The method is depended on the height of the 3D boxes. As such, the method of choice of the height values varies across different researchers and as discussed in the literature, some of the methods may not accurately determine the height.
 ii. The method cannot be used for rectangular images.
iii. In the z-direction of the images, the method tends to overcount the number of cubes covering the surfaces of the image. The method is also associated with overcounting and undercounting in the x- and y-directions. This may lead to distortion on the total number of boxes as computed by Equation (3.4).
 iv. The method uses linear least square regression to estimate the fit and the fractal dimensions of the data points on the log-log plots of N and 1/r. The method is associated with inaccuracies and does not take into account any nonlinearity of the data.

To overcome some of these challenges of DBC, several authors have suggested different approaches or modifications of the technique. For instance, Liu and others (2014) [12] suggested an improved DBC through modification of the box counting mechanism, shifting of box blocks in xy-plane, and choosing appropriate sizes of the grids. They illustrated that through these strategies, overcounting along the z-axis and undercounting boxes at the border of adjacent box blocks (where there is a sudden change of gray-level) were minimized. In a similar course, in their paper published in the *Journal of Measurement* [13], Panigraphy, Seal, and Mahata (2019) proposed four mechanisms of improving the DBC method: (1) a new formulation for box counting along z-axis to avoid overcounting of the boxes, (2) partitioning–shifting–partitioning approach to minimize the errors due to undercounting along the xy-directions, (3) formula to determine small values of the height of the boxes to reduce errors caused by using very large values of the height, and (4) an improved least square regression for accurate fitting of the data and determination of the fractal dimension. Hong, Pan, and Wu [14] proposed an improved DBC with the fractal characteristics treated as an eigenvector and when the method is combined with multiscale and multidirections, it is shown to be an effective tool for palmprint recognition. In a related study, Li, Du, and Sun [15] presented an improved DBC, which uses the smallest number of boxes to cover the whole surface of the image at each of the specific scales was proposed and it was shown to give accurate results as compared to the traditional DBC. Panigraphy et al. [16] have published three improved DBC methods based on the eigenvalue, kurtosis, and skewness of the grayscale image. In each of these methods, a new mechanism for xy-plane shifting and improved computation formula for the number of boxes required to cover grids on the xy-plane and weighted least square methods were used. These methods were shown to outperform the conventional DBC and some other improved DBC methods. There is a lot of exciting information on the enhancement of box counting method and the readers are referred to the published literature [3–5,17–22].

Strictly speaking and as it was stated earlier, the fractal theory is extensively used in literature, and the box counting method is the most common method of fractal analysis. Besides characterizing the fractal nature of general images, the box counting method has been evaluated and used for specific applications. For example, Xu, Jian, and Lian (2017) evaluated the use of phase fraction in the microstructure of Ni-B alloys [23]. The study revealed that the box counting method is an effective method of studying phase fraction in alloy compositions. In another study, Tripathy and others characterized the formation of porosity in Indian shales [24] and as illustrated in reference [25] and citations therein, it was concluded that box counting method is an effective procedure for studying porosity distribution in microstructures. The box counting method was also used by Tanaka, Kimura, Chouanine, Kato, and Taguchi in 2003 to study the fracture mechanism of the surfaces of metals and ceramics [26]. The fracture surfaces resulting from tensile and impact fracture (either brittle or ductile failure) have been shown to exhibit self-affine roughness [27]. It is stated in reference [27] that the self-affine characteristics of the fracture surface should be taken into consideration during fracture energy calculation. The box counting method was used to compute the fractal dimension of the fracture surface of cortical bone and a relationship was established between the fracture energy and the fractal behavior [28]. In thin film surfaces, box counting has been extensively used to compute the fractal dimension to explain the surface evolution during depositions [29]. In Chapters 4–6, detailed applications of the box counting method on fractal characterization, the fractal and parameter evolution, of thin films have been discussed.

3.3 TRIANGULATION METHOD

This technique is closely related to the box counting method and it uses pyramids rather than boxes. It is also known as the prism counting method and the procedure involves covering the area by a square patch and subsequently dividing it into triangles or imaginary prisms as shown in Figure 3.2 [30]. The procedure involves covering the area of the fractal feature by a square patch and then four triangles (one pyramid) are created by subdividing the square by joining the two diagonals of the square as shown in Figure 3.2a. Next, as shown in Figure 3.2b, the square area is further divided into four quadrants and each of the square four triangles are generated. This procedure is repeated until the length of each pyramid is equal to the resolution of the digital data of the image under consideration. The fractal dimension (D) in this method is then computed using the Equation (3.10).

$$D = 2 - \lim_{r \to 0} \frac{\log A(r)}{\log r} \tag{3.10}$$

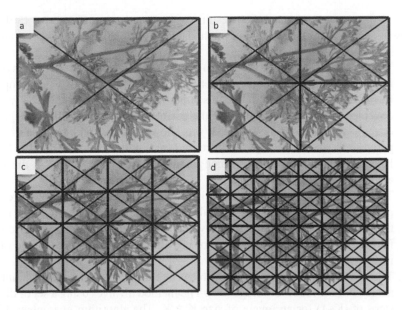

FIGURE 3.2 Triangulation method for computing the fractal dimension of a natural fractal feature.

where r is the base length of the prism and A is the apparent area of the prism. The slope of a double log plot of A versus r yields the fractal dimension (D) of the feature. The triangulation method is a very powerful method of computing D and the only challenge in using the method is in the determination of the number of steps and the corresponding size of the steps. As such, the focus of research has been on developing accurate sampling techniques for this fractal methodology. The original algorithm of the triangulation method was developed by Clarke in 1986 and it is known as the geometric-step technique [31]. The method is based on a series of geometric steps that increase by a factor of two until the maximum step as dictated by the algorithm. The maximum step size is usually determined by $(I - 1)/2$ where I represents the size of the image. If, for instance, the size of the image is 33 × 33, then the maximum step size would be 16 and that steps of 1, 2, 4, 8, and 16 may be used in the geometric-step method. The method ensures uniform distribution of points on the bi-logarithmic regression and that unbiased regression is achieved. However, when the image size (window within the image) is not of the pixels of order 2^n+1, there will be portions of the image which will not be covered by the geometric-step method and therefore considerable parts of the image will be omitted in the fractal estimation. For example, in an image with 29 × 29 pixels, only 17 × 17 pixels of the image will be covered if the geometric-step algorithm is applied. The limitation can be navigated by using the arithmetic-step method (1, 2, 3, 4, 5, …). The method yields sufficient points for the log-log regression and it can be used for even a small area of the image. However, the method may result into bias in the bi-logarithmic regression of the fractal dimension estimation. In their study, Ju and Lam in 2009, compared the coverage capability of various methods of sampling for the triangulation method (i.e. geometric-step [fixed coverage], geometric-step [varying coverage], arithmetic-step, and divisor-step methods) and their results are summarized in Figure 3.3 [30]. The following conclusions about the four methods discussed in that study were drawn.

i. The geometric-step and fixed coverage method results in 'wasted' pixels of the image since the coverage is not very satisfactory and as such the computed fractal dimension is not a true representative of the entire window. The varying coverage method allows the triangular prisms to expand within the image and cover a wider area of the image. It was shown to cover a larger area as compared to the fixed coverage as illustrated by Figure 3.3 (a1–a3) and (b1–b3). However, the two variations do not guarantee complete coverage of the study area represented by the image.

ii. The arithmetic-step method allows the use of a sufficient number of regression points, especially when using a small image. As shown in Figure 3.3 (c1–c3), the method improves on the area coverage as compared to the geometric-step. However, it can be seen that it is not possible to achieve full coverage of the image window.

iii. The divisor-step method proposed by Ju and Lam was shown to guarantee a nearly 100% window coverage [30]. The method involves using a set of divisor steps of $(K-1)$ for an image of size $K \times K$. The algorithm guarantees 100% coverage of the image window as it can be seen in Figure 3.3(d1–d3). In this way, the computed fractal dimension would be an accurate representation of the image under consideration. Ju and Lam have detailed the application and use of the divisor-step methodology in determining fractal dimension and readers are referred to their article [30].

From the preceding discussion, three significant factors should be considered in a triangulation prism method, namely, the number of steps, the size of the step, and the estimation accuracy of the area under consideration. These factors were described in detail by Zhou, Fung, and Leung (2016) [32]. It was discussed that the higher the number of steps, the better the accuracy of computation of the fractal dimension. It was also discussed that the step size should have an even logarithmic scale and should be carefully selected to achieve sufficient coverage of the window [30,31]. A lot of work currently being undertaken in this method is on the improvement of the surface coverage of the image for better estimation and representation of the fractal dimension.

3.4 BROWNIAN MOTION METHODS

Brownian motion is defined by random movement of matter which is generated through collisions among the particles of the matter. In terms of Brownian motion, fractal dimensions (and other fractal characteristics) of surfaces and features are determined on the assumption that the features exhibit random processes with non-independent and normally distributed increments [6,33]. In fact, non-stationarity and nondifferentiability are the two main properties of Brownian motion; however, for fractal Brownian motion, besides these two properties, it has a statistical dependence. This statistical dependence is measured by a Hurst exponent (H). Fractal Brownian

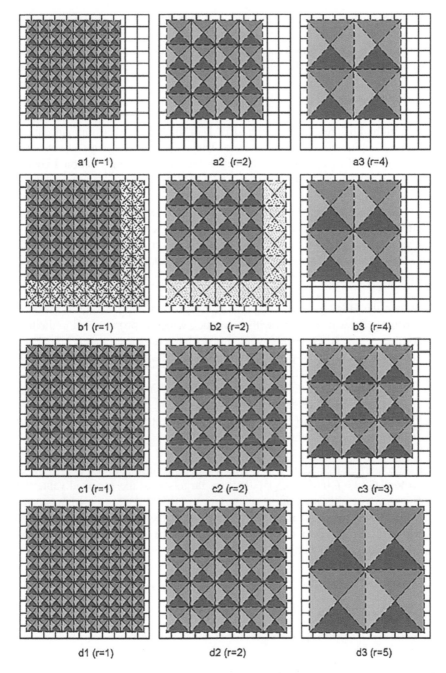

FIGURE 3.3 Demonstrating the coverage capability of different sampling methods for triangulation technique on an image of size 11×11 pixels. The images represent the first three steps for (a1–a3) geometric-step method (fixed coverage) (b1–b3) geometric-step method (varying coverage); (c1–c3) arithmetic-step method; (d1–d3) divisor-step method (image reused with permission from Elsevier Ltd. [30]).

motion (abbreviated as fBM and denoted as $B_t{}^H$) is defined as a continuous and random process (t) in a range of non-negative and real values [34]. The variance of an fBM process has a power law relationship with the Hurst exponent (H), which takes any value between 0 and 1 (Equation 3.11).

$$Var\left(B_t^H\right) \propto t^{2H} \tag{3.11}$$

The definition of discrete time version (dfBM) of the fBM is necessary since most of the practical engineering data are discrete. The series of increments of consecutive dfBM, determined as $B_i^H - B_{i-1}^H$, is known as the discrete-time fractional Gaussian noise, abbreviated as dfG_n and denoted as G_i^H [35]. From these definitions, the sum of dfG_n gives the dfBM and each series of dfBM has a relationship with dfG_n. Additionally, the two variables exhibit the same Hurst exponent (H). However, while the dfBM is related to nonstationary processes, dfG_n exhibits stationary mean and variance over the span of time. Based on these descriptions, the following three conditions exist and are very useful during the fractal characterization of surfaces (Brownian motion).

i. When the Hurst exponent (H) is 0.5, the process (B_i^H) is said to exhibit the ordinary Brownian motion with a variance which is proportional to the length of the series, and G_i^H exhibit white noise characteristics (random signal at any time and the values of power in a log-log power spectrum do not change with the frequency of the signal).

ii. When H is less than 0.5, the B_i^H is said to exhibit a sub-diffusive motion and the G_i^H are anti-persistent (negatively correlated).

iii. In cases where H is larger than 0.5, the B_i^H is said to be over-diffusive while the G_i^H is positively correlated (or persistent).

As stated earlier, $B_i^H = G_1^H + G_2^H + \ldots + G_i^H$, and defining the variance of the dfG_n as $\sigma^2 = Var\left(G_i^H\right) = E\left[(B_i^H - B_{i-1}^H)^2\right]$, then for large series, the expected value (E) of dfBM is zero. The self-similarity characteristics of dfBM are expressed in two ways: its variance as a power function of the length (n) of the sample and autocorrelation function of dfG_n. The expressions for the two cases are written as follows (Equation 3.12 and Equation 3.13) without derivation.

$$Var\left(B_n^H\right) = Var\left\{B_0^H, B_1^H, \ldots, B_n^H\right\} = Cn^{2H} \tag{3.12}$$

$$\rho dfG_n\left(k\right) = \frac{1}{2}\left(|k+1|^{2H} + |k-1|^{2H} - 2|k|^{2H}\right) \tag{3.13}$$

In general, the expected covariance (E) of a continuous time fractal Brownian motion series between two times, t and T is given as shown in Equation (3.14).

$$E\left[B_t^H B_T^H\right] = \frac{\sigma^2}{2}\left[t^{2H} + T^{2H} - |T-t|^{2H}\right] \tag{3.14}$$

From Equation (3.14), the study [34] derived an expression for auto-correlation function for dfBM series and it was written as shown in Equation (3.15).

$$\rho dfBM(k) = 1 - \frac{n(n-1)k^{2H}}{2\sum_{i=1}^{n-1}\left[(n-i)i^{2H}\right]} \tag{3.15}$$

These formulations form the basis for fractal computations for Brownian motion processes. In image analyses, the surface profiles exhibit random and nonstationary characteristics and can therefore be considered as fractal Brownian motion processes [36]. There are two common methods based on Brownian motion formulations for computation of fractal dimensions; namely, power spectrum and variogram methods [6,37]. In an overview presented in the chapter by Mwema et al. [6], both of these methods were discussed and their underlying principles illustrated and the readers are referred to the reference for more information. In this section, a discussion on the power spectrum method is briefly extended since it has been used in Chapter 5 for determining the fractal dimension of the sputtered thin films.

POWER SPECTRUM METHOD

This method utilizes Fourier transforms to decompose an image into waves of different wavelengths and frequencies. Mostly, a two-dimensional fast Fourier transform (2D-FFT) algorithm is used to transform the topography images (information) into the complex domain. A general expression for 2D power spectral density function (2D-PSDF) of the surface whose topography is described by z(x,y) can be represented as shown in Equation (3.16) [38,39].

$$PSD(f_x, f_y) = \lim_{L \to \infty} \frac{1}{L^2} \left\{ \int_{-\frac{1}{2}L}^{\frac{1}{2L}} dx \int_{-\frac{1}{2}L}^{\frac{1}{2L}} dy. z(x,y) \exp\left[2\pi j(xf_x + yf_y)\right] \right\}^2 \tag{3.16}$$

In Equation (3.16), L is the length of the window of the image (usually assumed to be equal in x- and y-directions), f_x, f_y represent the spatial frequencies in the x- and y-axes, z (x,y) is the profile of the topography (height surface roughness), and j is a complex number. Mathematically, Equation (3.16) requires further definitions to be applied in the computation of the FFT and finite values and boundary conditions should be defined for a specific application [38].

In the surface analysis, the PSD is usually computed and plotted against the spatial frequencies (f) on a double logarithmic scale. The nature of the PSD profiles depends on the surface roughness of the surface. For example, it was discussed by Mwema et al. [40] that surfaces with random height feature exhibit a plateau at the low spatial frequency and an inverse slope at high frequencies with a visible knee at the transition region (Figure 3.4). It was also described that surfaces with periodic features exhibit PSD profiles with multiple peaks at the region of low frequency and a monotonically

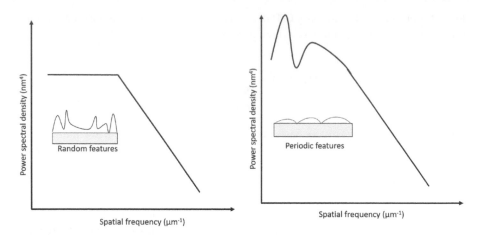

FIGURE 3.4 Typical PSD profiles for surfaces with random and periodic features.

decreasing PSD at higher spatial frequencies (Figure 3.4). The mathematical descriptions of the PSD profiles for topography can be undertaken by fitting the PSD data into known models. The authors of this book have utilized two common models namely k-correlation (also known as the ABC) model and power law function in PSD analyses of surfaces of aluminium thin films [38,41,42]. The expression representing the k-correlation model is shown in Equation (3.17).

$$PSD = \frac{A}{\left(1 + B^2 f^2\right)^{\left(\frac{C+1}{2}\right)}} \tag{3.17}$$

Where A, B, and C are constants of the k-correlation model and each of them have physical meaning to the topography of a surface. Constant A is known as the shoulder parameter and it represents the value of the power spectrum at very low spatial frequency. In this region, the PSD profile is nearly flat, depicting a white noise condition explained by the fBM general formulation earlier. For an AFM image analysis of a surface of a sample, large values of A indicate the presence of peaks and large surface structures. The constant B describes the transition region between high and low spatial frequencies of the PSD profile. It shows the value of the knee and its in-plane correlation length and grain size. It is also known as the correlation length of a surface and it is used to show the lateral variation of surface roughness on a surface. Generally, the smaller the value of constant B, the larger the lateral roughness of the surface. The PSD profile at high spatial frequencies is described by the constant C, which is associated with the high correlation behavior of the surface roughness. For purely fractal surfaces such as glass, C is usually 3. It should be noted that the k-correlation model is used to describe the PSD profile only up to the transition zone beyond which the inverse power law can be used. The expression for the inverse power law describing the PSD profiles at high spatial frequencies (related to the autocorrelation function of fBM) is shown in Equation (3.18). In the equation, K is known as the spectral length and α is the gradient of the PSD profile.

$$PSD = \frac{K}{f^{\alpha}} \qquad (3.18)$$

Based on the power law, the fractal dimension, D, can be determined as follows. When $0 \leq \gamma < 1$, D = 2, or $3 < \gamma$, then D = 1; otherwise $D = \frac{1}{2}(8 - \gamma)$. The Hurst exponent (H) can also be determined and it is related to the fractal dimension as follows; D = 3 − H.

The procedure for implementing the PSD algorithm and interpretation is stepwise and can be undertaken in any simple programming software. For instance, Mwema et al. [42] analyzed the fractal characteristics of Al thin films using the power spectrum algorithm implemented in a simple MATLAB® code. It was discussed that before MATLAB programming, the surface topography of the image should be carefully obtained with minimum artifacts and when necessary, the artifacts may be corrected. Details on achieving quality images for AFM microscopy were described in an earlier publication by the authors and it was shown that proper choice of the AFM parameters or correction of artifacts from the image can considerably improve the quality of the surface topography data [43]. It is also important to expose all the images to the same pre-processing procedure to avoid abnormalities in the fractal results.

In describing surfaces such as those of thin films using PSDFs, the following important points are key to remember.

i. The fast Fourier transform algorithm decomposes the topography image into waves of different frequencies and wavelengths. A PSD analysis is undertaken on each wavelength to generate a waveform image, known as the power spectrum as shown in Figure 3.5(a). As shown, the spectrum image consists of two regions, namely, the central region and the edge (Figure 3.5(b)). The central region is usually white (by default) while the edges are usually dark. The central region represents the region of low frequencies whereas the edges represent the regions of high frequency.

ii. The interpretation of the power spectrum (wave images are shown in Figure 3.5) is necessary for complete fractal analysis of a surface using the PSD method. The size and shape of the white central patch is one of the descriptive factors for the lateral roughness of the images [44]. A very large central region indicates the presence of generally large surface features such as grain sizes, porosity, and phases on the microstructure. Additionally, the physical transformation of the shape of the white feature could be an indication of contrast in lateral evolution especially due to change in processing parameters of a surface. The presence of a very bright central white feature on the power spectrum is an indication of the domination of horizontally oriented surface structures on the microstructure. Additionally, if the power spectrum image is dominated by either a dark or white region, it means that the surface topography has very high lateral homogeneity. In some instances, white lines may be observed

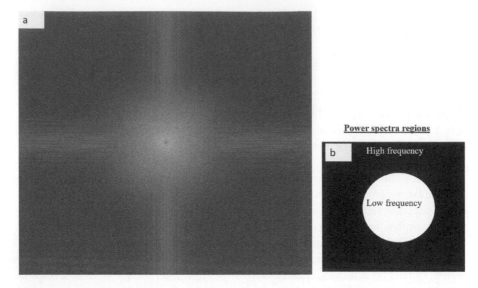

FIGURE 3.5 (a) A typical power spectrum of an image for a surface topography and (b) schematic representation of the power spectrum regions.

within the dark regions of the power spectrum; such observations indicate the presence of lateral spaces along the surface of the microstructure.

iii. The PSD profiles, which are usually plots of power spectrum versus the spatial frequencies on a log-log scale, provide important physical characteristics for description of the surface topography even before curve fitting into the theoretical models. As shown in Figure 3.4 above, the PSD profiles, generally, exhibit three regions, namely (1) the low spatial frequency and high-PSD region, (2) the transition region, and (3) high spatial frequency regions. The shape and size of different regions have indicative meanings to the surface topography of a surface. For instance, Mwema et al. (2018) [38] discussed that Al thin films with a larger white noise region indicate that the surfaces have the homogenous lateral distribution of roughness of the structures [38]. In fact, surfaces with highly serrated line profiles were shown to exhibit a smaller white noise region. The size of the logarithm X-scale is also another indicator of the fractal characteristics of a surface. It was noted in the same study [38] that surfaces with a high order of the x-scale (for PSD profiles) have smaller spatial wavelengths and are indicative of the homogenous distribution of small particles over the surface of the microstructure.

iv. Another important consideration in PSD profiles is the region of high correlation (power law region). If the profile is dominated by the power law, it means that the topography of the sample represented by the image is generally rough. These results have an implication on the choice of AFM parameters for the analysis of surface topography. For PSD profiles dominated by power law region, it is advisable to use a scan size of about 3 × 3 μm or smaller as it was demonstrated by Mwema et al. (2020) on their paper titled 'Effect of Scan Size on the Scaling Law of Sputtered Aluminium Thin Films' [45]. On the contrary, if the profiles are dominated by k-correlation model, it means that the surfaces have less sharp and periodic features.

v. The curve fitting parameters for the power law (Equation 3.18)-exponent (α) and PSD amplitude (K) are interpreted as follows. The exponent is a measure of the evolution of the surface structures. For example, when Al thin films were deposited at higher sputtering power [42], their power law exponent was higher than those deposited at lower sputtering power and this is because power contributes to the evolution of surface features in thin films [46,47].

vi. The interpretation of the curve fitting parameters in the k-correlation model can be based on the mathematical meaning of the constants. (a) Parameter A is directly related to the central white region of the waveform image (Figure 3.5). The larger the value of A, the larger the circular patch on the spectrum image. As stated earlier, large values of A (larger radius of the white region) indicate larger morphological features of the surface. (b) The parameter B is the correlation length and indicates the transition from peaks to valleys and vice versa [42]. The larger the values of B, the lower the lateral roughness of the topography represented by the image. (c) The parameter C describes the finest details of the microstructure and it is associated with the dark regions of the waveform image. It has been discussed in the literature that C shows the sizes of the surface structures. The parameter (C) can explain the relationship between the growth of surface features and the conditions of the physical process. A large value of C is an indication of a surface with well-defined grains and large structures. In fact, C can be used to indicate an occurrence of nucleation and grain growth on a surface.

vii. From k-correlation model, further parameters, equivalent roughness (usually denoted as σ) and correlation length (τ) can be determined as $\sigma = \dfrac{2\pi A}{B^2(C-1)}$ and $\tau = \dfrac{(C-1)^2 B^2}{2\pi^2 C}$, respectively. These two parameters are also indicators of growth of the surface structures and their increase means that the surface features of the microstructure are growing. As such, the parameters can be used to describe the influence of various growth parameters on the surface evolution.

viii. The inverse power law results on the fractal dimension (D) and Hurst exponent are the final outputs of a PSD methodology. The interpretation of H in respect to the surface topography of micrograph images is based on the fBM criterion discussed earlier on the general Brownian motion formulations.

It is of no doubt that PSDF has been used in the characterization of surfaces and a lot of literature is available on this subject [29,38,39,48–51]. The general inference from these publications is that the accuracy of the fractal characterization depends on the quality of the image/signal, image processing, and curve fitting. As such, the user of the PSD method should have sufficient expertise in these different areas. It is important to note that PSD has been a very successful method for undertaking fractal characterization of pure thin films and the authors of this book have published several works (such as [38,42,47]) on PSD characterization on pure Al thin films grown via RF magnetron sputtering technology.

3.5 MULTIFRACTAL THEORY

3.5.1 BASIC FORMULATIONS

In most cases, the surfaces of materials cannot be described by a single scaling exponent. In such cases, the surfaces are said to be multifractal, and multifractal theory deals with the measurement of probability functions for surface features. The theory of multifractals is a very powerful tool in providing information on lateral (roughness) characteristics on the surface morphology of surfaces of materials and their growth mechanisms. The standard multifractal formulation is based on the assumption that the surfaces are self-similar and measures of the Euclidean space.

In this section, the theoretical formulation of the multifractal theory is presented based on the extensive publications in literature [52,53]. If μ is the measure on a Euclidean space of dimension E and that in x-space, μ can scale with different exponents defined as $\alpha(x)$, then,

$$\mu\left(B_\varepsilon\left(x\right)\right) \sim \varepsilon^{\alpha(x)} \tag{3.19}$$

Assuming that the formulation is based on the box counting technique, $B_\varepsilon(x)$ represents a grid (or box) of length ε. From Equation (3.19), $\alpha(x)$ is defined as the singularity index or singularity strength and it is defined by Equation (3.20).

$$\alpha\left(x\right) = \lim_{\varepsilon \to 0} \frac{\log \mu\left(B_\varepsilon\left(x\right)\right)}{\log \varepsilon} \tag{3.20}$$

Mathematically, the $\alpha(x)$ is defined as the local scaling exponent of μ at $x \in E$ (E is the topological space). Yadav et al. [53] stated that in practical applications, the limit $\varepsilon \to 0$ does not have any significant meaning, and as such it is meaningful to define the exponent in terms of coarse Hölder exponent. For instance, Equation (3.21) describes the coarse Hölder exponent of sub-square Q as it was illustrated by Yadav et al. and many other literatures [53,54].

$$\alpha\left(Q\right) = \frac{\log \mu\left(Q\right)}{\log \varepsilon} \tag{3.21}$$

Following Equation (3.21), the concept of the multifractal spectrum ($f(\alpha)$) can be derived. If the space is divided into grids of size ε and the number of grids corresponding to these sizes are denoted as $N_\varepsilon(\alpha)$, then the multifractal spectrum can be defined as

$$f_\varepsilon\left(\alpha\right) = -\frac{\log N\varepsilon\left(\alpha\right)}{\log \varepsilon} \tag{3.22}$$

The multifractal spectrum function $f(\alpha)$ represents the fractal dimension of the subsets of grids of size ε units with a coarse Hölder exponent (local scaling indices) of α. It is important to note that it is not possible to directly determine the variables in Equation (3.22) because the logarithm of zero is infinite and usually the determination can be based on a log-log plot at different values of ε. As the values of ε become too small (i.e. ε→0), the values of $f_\varepsilon(\alpha)$ tend to a well-defined mathematical limit. It, therefore, means that for very small values of ε there are $\varepsilon^{-f(\alpha)}d\alpha$ which have values ranging between ε^α and $\varepsilon^{\alpha+d\alpha}$.

Yadav et al. [53] described two methods of determining the spectrum function f(α), namely method of moments and histogram method. The histogram method for determining the multifractal spectrum function involves the six key steps. Considering the Euclidean space of μ, then the following steps apply for the method of a histogram.

i. The set defined by Q is covered by grids of defined size (ε) and the correspond-ing number of grids is then determined as $N_\varepsilon(\alpha)$.

ii. Then, for each set of Qs (each box), the coarse Hölder exponent α_i is deter-mined using Equation (3.21). The space μ_i can be defined as the measure for the grid *i*. As such the Equation (3.21) can be rewritten as $\alpha_i(Q) = \dfrac{\log \mu_i(Q)}{\log \varepsilon}$.

iii. Plot a histogram for the value α computed in the procedure (ii) with the bin size of Δα. Then, the number of densities $N_\varepsilon(\alpha)$ is computed such that the number of sets which have coarse Hölder exponent between α and α+Δα is $N_\varepsilon(\alpha)\Delta\alpha$.

iv. Using different values of ε, steps (i) to (iii) are repeated,

v. Using Equation (3.22), the plots of $\dfrac{\log N\varepsilon(\alpha)}{\log \varepsilon}$ against α are generated for the different values of ε.

vi. The function f(α) is estimated from the slope of the plot generated in step (v). The parameter μ is known as a multifractal measure if all the curves converge to a single curve f(α) as the size of the grids (ε) becomes small.

The histogram method has been used in various cases, for example, it was used by Franca et al. [55] to study the brain dynamics, by Drzewiecki et al. [56] to describe the satellite image content, and so forth.

The second method for determining the multifractal spectrum is known as the moment's method. The method described here was adopted by the authors in an article on the characterization of the fractal nature of hillocks on pure Al thin films grown via RF magnetron sputtering [57]. If we consider that ε lies between 1 and L (the length scale and upper bound on the scale, respectively [58]), then the probability of distribution P_{ij} for the gray values in the boxes (i,j) can be determined as shown in Equation (3.23).

$$P_{ij}(\varepsilon) = \frac{n_{ij}}{\sum n_{ij}} \qquad (3.23)$$

From Equation (3.23), the n_{ij} represents the average gray values of the boxes of length ε. The probability distribution and the length of the box within the said ranges are related by the singularity exponent described earlier in this section. The relationship can be written as shown in Equation (3.6).

$$P_{ij}(\varepsilon) \propto \varepsilon^\alpha \qquad (3.24)$$

As shown in the equation, the singularity exponent is the subset of probabilities of distribution and depends on the grid (i,j). It is important to note that when all the grids on the Euclidean space have the same value of singularity exponent/strength, the surface is said to exhibit a mono-fractal character, otherwise, a multifractal behavior exists. Mathematically, Equation (3.22) can be rewritten as shown in Equation (3.25).

$$N\left(\varepsilon\right) \propto \varepsilon^{-f(\alpha)} \tag{3.25}$$

In this case, the singularity spectrum or multifractal spectrum is said to be a continuous function of the singularity exponent. For a given space partitioned into several grids/boxes, a partition function denoted as Z(q,ε) can be defined as shown in Equation (3.26).

$$Z\left(q,\varepsilon\right) = \sum_{i=1}^{N(\varepsilon)} \left\{\mu Q_i\right\}^q = \sum_{i=1}^{N(\varepsilon)} \left\{\varepsilon^{\alpha(Q_i)}\right\}^q = \sum P_{ij}^q = \varepsilon^{\tau(q)} \tag{3.26}$$

It can be seen that this parameter is a function of the size of the boxes and another parameter denoted as q. In this case, q is known as a moment order and it is a real number varying between $-\infty$ and $+\infty$. The parameter $\tau(q)$ is known as the mass or correlation exponent and it be computed from the gradient of the bi-logarithmic plots of $\ln[Z(q,\varepsilon)]$ against $\ln[\varepsilon]$. The relationship between the multifractal spectrum function and the mass exponent is represented in Equation (3.27). The derivation of the relationship is based on Legendre transform of the partition function.

$$f\left(\alpha\left(q\right)\right) = \alpha\left(q\right)q - \tau\left(q\right) = q\frac{d\tau\left(q\right)}{dq} - \tau\left(q\right) \tag{3.27}$$

The mass exponent can be determined using the following relationship (Equation 3.28).

$$\tau\left(q\right) = q\alpha\left(q\right) - f\left(\alpha\left(q\right)\right) \tag{3.28}$$

From the above formulations, the following points are important to remember.

i. The partition function Z(q,ε) is determined using the Equation (3.22) based on the experimental data on the computation of μQ.
ii. The double logarithmic plots of Z(q,ε) against ε is usually a straight line whose slope represents the mass exponent τ(q) based on Equation (3.26). It is important to note that in real world experimental data, the power law scaling as depicted by the equation is not usually true especially for very large or small values of ε. It was however explained by Yadav et al. and others [53,59].
iii. From the slope, the values of τ(q) and f(α) can be determined using Equation (3.10).
iv. A plot of f(α(q)) against the α(q) is known as the multifractal spectrum. It shows the different values of fractal dimensions and the probability of

distributions. The shape and width of the plots are used to describe the multi-fractal behavior of the surfaces and evolution of physical processes such as grain formation and surface defects [60]. The function is a powerful method for expressing variability in the scaling properties of measures such as contents of carbon and clay in a soil sample.

v. The generalized fractal dimension D(q) is directed related to the mass expo-nent as shown $D(q) = \dfrac{\tau(q)}{q-1}$. A plot of D(q) against q is usually used in charac-terizing the multifractal behavior of surfaces. For a multifractal surface, the D(q) decreases nonlinearly leading and exhibits important parameters as it will be illustrated in Chapter 6.

In implementing the multifractal theory for examining soil spatial variation, Biswas, Cresswell, and Bing (2012) [61] described the following steps.

i. The first step is to compute the measure of probability (P) of the space (in this case it may be the linear distance from a rectangle for an area, etc.). Then, the number of samples for the analysis (the number of grids for example) are determined. The sufficient number of minimum samples for the multifractal analysis is very important and should be determined to save on time and cost of computation. The determination of this number depends on the possible subdivisions of the space and the statistical moment's range.

ii. The next step is to evaluate the multifractality of the data using the parti-tion function (Z (q,ε)). The function should be plotted on a bi-logarithmic graph against the size of the subdivisions (ε). The function should exhibit a linear relationship with the distance for multifractal behavior, otherwise, the data are mono-fractal. If the data are multifractal, the next step should be implemented.

iii. Carry out a linear regression on the plot of the partition function against the size and determine the mass exponent $\tau(q)$ as the intercept of the plot.

iv. Next, from the linear regression of the plot of $f(q,\varepsilon)$ versus $\log(\varepsilon)$, determine f(q), α(q), D(q) as intercepts of the plots.

v. Finally, generate the plots of mass exponent ($\tau(q)$) as a function of moment order, (q), f(q) as a function of α(q), and D(q) as a function of q.

As can be seen, the multifractal analysis is a step-by-step process and therefore a simple computer program can be written for the formulations presented above. Computer software such as MATLB, SAS, and Mathcad can be used to implement these steps. Additionally, image analyses software such as Fiji (ImageJ) have plug-in (e.g. FracLac) which have proven to be effective tools in multifractal analysis. It is important to understand that the above multifractal description can be used only for a single variable along with its spatial support. The approach may be limited where complex systems result in long-range cross-correlated, multifractal, time series. In such cases, *a joint multifractal analysis* is utilized. By definition, joint multifractal analysis is used in determining multiscale spatial relationships between two or more variables along with common spatial support. Similar to multifractal analysis, the joint multifractal method involves dividing the system into small parts of size ε.

Then, the probabilities of measure for both variables in the ith part (segment) are defined. It should be noted that for each variable, a corresponding singularity strength and moment order shall be defined, and therefore for a two-variable multifractal characterization, two singularity strengths and moment orders are defined, and so forth. Since joint multifractal analysis is not extensively used in microstructural characterization, the mathematical formulations have not been presented in this book. The readers are, however, referred to the literature [62–64] and it can be noted that most of the joint multifractal analyses are employed in natural sciences such as agronomy, crop science, soil science, forestry, and so forth [65]. It also finds applications in analyzing financial market data variables [66,67], transportation, medicine, and geophysics. The joint multifractal analysis uses a joint partition function and it is a very important tool in comparing the correlations among the variables under consideration. By varying the values of moment order for each variable, it is possible to obtain different pairs of the singularity strengths and Pearson correlation analysis can then be used to quantify the relationship between the singularity strengths (scaling exponents) of the different variables. From the Pearson correlations, it is possible to evaluate the strength of the correlation between the variables – whether strong or weak. Similar to the multifractal analysis for a single variable, the joint partition approach is a stepwise method and can be implemented in a computer program [62].

3.5.2 Multifractal Detrended Fluctuation Analysis (MFDA)

To study the multifractal nature of non-stationary time series, a multifractal detrended fluctuation analysis (MDFA) is used [68–72]. To illustrate the formulation of the one-dimensional MDFA, we denote a time series by an array: $Y(j)$, where $j = 1, 2, 3, \ldots, M$. Suppose the time series is divided into parts of equal length, m, then the total number of parts is given by $M_m \equiv \text{int}[M/m]$. It means that the smaller the divisions (parts), m, the larger the number of parts M_m. If the different parts are denoted by P_n, where $n = 1, 2, \ldots, M_m$, then the value of the time series at an arbitrary point defined as jth point of the subdivisions, P_n, is given as $Y_n(j)$, and can be determined as $Y_n(j)=Y((n-1)m+j)$. For each part of the time series, P_n, the cumulative sum of the value of the time series can be given as shown in Equation (3.29).

$$S_n(j) = \sum_{k=1}^{j} P_n(k) \tag{3.29}$$

The cumulative sum $S_n(j)$ is predicted to be a polynomial function (defined as $\check{S}_n(j)$) and its coefficients can be obtained using the least square method. As such, the root means square (rms) fluctuation from the polynomial characteristic which is normally denoted as F (n,m) is given by the Equation (3.30).

$$F^2(n,m) = \frac{1}{m} \sum_{j=1}^{m} \left\{ S_n(j) - \check{S}_n(j) \right\}^2 \tag{3.30}$$

Therefore, the qth-order fluctuation function of the Equation (3.30) can be given as shown generally by Equation (3.31).

$$F_q(m) \equiv \left\{ \frac{1}{M_m} \sum_{n=1}^{M_m} \left[F(n,m) \right]^q \right\}^{1/q}$$
(3.31)

From Equation (3.31), it is postulated that q can take any real value, and therefore for a multifractal time series, the following relationship (Equation 3.32) between rms fluctuation and q can be derived.

$$F_q(m) \sim m^{h(q)}$$
(3.32)

From Equation (3.32), a plot of $F_q(m)$ against m can be generated on a double logarithmic scale. For a multifractal behavior, the plots consist of linear and nonlinear sections. The exponential values ($h(q)$) are referred to as generalized Hurst exponents and can be obtained as the slope of linear sections of the $\log(F_q(m))$ versus $\log(m)$. The $h(q)$ relates to the mass exponent $\tau(q)$, moment order (q), and multifractal spectrum as illustrated in Equations (3.33) and (3.34). These equations show that the multifractal spectrum that characterizes detrended fluctuations in a time series can be obtained from the $h(q)$.

$$\alpha(q) = h(q) + qh'(q)$$
(3.33)

$$f(\alpha) = q \left[\alpha - h(q) \right] + 1$$
(3.34)

The formulations above apply to one-dimensional MFDFA. However, there is a lot of interest on the two-dimensional MFDFA formulations and it was initially proposed by Gu and Zhou in 2006 in a powerful article published in the esteemed journal of *Physical Review E* [73]. The Equations (3.32), (3.33), and (3.34) for the 2D case can be written as follows.

$$F_q(m) \equiv \left\{ \frac{1}{M_m \times L_m} \sum_{a=1}^{M_m} \sum_{b=1}^{L_m} \left[F(a,b,m) \right]^q \right\}^{1/q}$$
(3.35)

$$\alpha(q) = h(q) + qh'(q)$$
(3.36)

$$f(\alpha) = q \left[\alpha - h(q) \right] + 2$$
(3.37)

The above equations have been derived based on 2D array in i- and j-directions and the 2D height space are divided into M_m and L_m disjoint square segments. The determination of the generalized exponents is similar to the 1D case and can be obtained from the linear plots of the $F_q(n)$ plots.

3.5.3 APPLICATIONS OF THE MULTIFRACTAL THEORY

It is no doubt that a multifractal characterization is a powerful tool in various fields. A lot of publications are available on the analysis of multifractal problems in different fields. Herein, we illustrate a few examples of multifractal characterizations based on the existing literature for various fields. However, it is important to note that detailed studies on the multifractal characterization of thin films shall not be presented here since the entire Chapter 6 is dedicated for such. The application of the multifractal spans from materials characterizations to financial and market analyses, geotechnical engineering to medical sciences, and so forth.

   i. In finance and markets, MFDA is used in studying the fluctuations characteristics of the prices in the stock market. An MFDA study on the NASDAQ composite index showed that the stock market is influenced by both large and small fluctuations and it is not a random process [74]. Oral and Unai used MFDA to model and forecast time series for price fluctuations of gold, silver, platinum, and other precious metals [75]. Agterberg has illustrated the use of multifractal analysis in modeling the metal size–frequency distributions in Canada and the world [76]. The price index correlations for different prices over a long period can be studied via multifractal methods [77]. More studies are available on the multifractal characterization of agricultural markets [78–81].
  ii. In geology, multifractal methods can be used to study the distribution of geochemical elements in a specific region [82]. It is also illustrated by Munoz et al. [83] that multifractal techniques can be used to characterize the distributions of landslides. Multifractal characterizations can be used to study the variations in the enrichment of geotechnical fields [84]. It has been shown that geophysical data such as electrical conductivity and magnetic data have multifractal characteristics and as such multifractal methods can be used to study geophysical data [85].
 iii. In the medical field, the multifractal method can be used to quantitatively classify cell types in humans and animals [86]. West and Scafetta studied the dynamics of human gait through multifractal characterizations [87]. In a similar approach, the multifractal dynamics of heart rates can be characterized and classifications of heart dynamics is possible [87]. The method has also been used to detect abnormalities in the backbone structure [88], assessing cancer risk [37,89,90], and heterogeneity in biological tissues [91].
  iv. In soil science, the multifractal method can be used to describe the heterogeneous properties and variability in of soils with different textures [92], characterization of soil pore characteristics [93–96], soil moisture time series [97], and modeling of the microtopography of soils with multiple transects data [97]. The multifractal methods can also be used to characterize particle-size distributions (PSD) of soils [98]. The joint multifractal method has been used to study the influence of topography and texture of soils on their water storage capacity [99].

3.6 OTHER METHODS

There are exhaustively so many other methods for studying the fractal behavior of systems other than those discussed in this chapter. The methods detailed in this chapter are skewed toward surface topography characterization. Some of the other methods of fractal characterization include area-based methods, correlation functions, and Minskowski functionals [29]. The area-based methods are related to triangulation and box counting technique, although they also include techniques of relating area and perimeter of the surface structures as described in the pioneering works by Mandelbrot[100–102] and other publications [57,103–105]. The other area-based techniques include a blanket, isarithm, and Korcak's methods [2,106]. The correlation methods are derived from the fractal Brownian motion theory and the most common are the autocorrelation and height–height correlation functions. These functions have been described in reference (and the citations therein) [29] and others [107,108], and applied in Chapters 5 and 7 of this book in fractal analyses of the surfaces of Al thin films. Finally, the Minkowski functionals are also used in the fractal characterization of surfaces and are based on the segmentation principle of separating a topographic (or any other image) into two parts, that is, high and low parts [109]. There are three Minkowski functionals, namely, the volume, boundary length, and Euler characteristic/connectivity and have been described in various publications [110–115] and used in the fractal analysis in Chapters 5 and 7 of this book.

3.7 SUMMARY

In this chapter, descriptions of methods for fractal and multifractal characterizations have been presented. The main objective for fractal analysis is to compute the fractal dimension, which is extensively used in describing the fractal properties. The most common method for fractal characterizations especially for images of microstructures is the box counting technique. The method is simple and versatile and has extensively been studied for the improvement of its accuracy. Other methods include triangulation and other area-based methods, PSD analysis, and other Brownian motion methods. A detailed theory on formulations of multifractal algorithms has been presented and the box counting method forms the basis of these formulations. General applications of multifractal theory in fields other than thin film characterization have also been presented since they are not readily available in a single resource. In each of the described methods, basic formulations, implementation steps, and key aspects of interpretation for surface analyses are explained. It is imperative to note that the methods of fractal measurements are inexhaustible and there are different versions of existing methods. Additionally, new methods are being explored and developed. In the subsequent sections of the book, Chapters 4–7, case studies based on authors' experimental works and published literature on the applications of these techniques have been illustrated.

REFERENCES

[1] B. Klinkenberg, "A review of methods used to determine the fractal dimension of linear features," *Math. Geol.*, vol. 26, no. 1, pp. 23–46, Jan. 1994.

[2] S. R. Nayak, J. Mishra, and G. Palai, "Analysing roughness of surface through fractal dimension: A review," *Image Vis. Comput.*, vol. 89, pp. 21–34, Sep. 2019.

[3] S. R. Nayak, J. Mishra, and G. Palai, "An extended DBC approach by using maximum Euclidian distance for fractal dimension of color images," *Optik (Stuttg).*, vol. 166, pp. 110–115, 2018.

[4] S. R. Nayak and J. Mishra, "An improved method to estimate the fractal dimension of colour images," *Perspect. Sci.*, vol. 8, pp. 412–416, Sep. 2016.

[5] S. R. Nayak, J. Mishra, A. Khandual, and G. Palai, "Fractal dimension of RGB color images," *Optik (Stuttg).*, vol. 162, pp. 196–205, 2018.

[6] F. M. Mwema, E. T. Akinlabi, O. P. Oladijo, O. S. Fatoba, S. A. Akinlabi, and S. Tălu, "Advances in manufacturing analysis: Fractal theory in modern manufacturing," in *Modern Manufacturing Processes*, First., K. Kumar and J. P. Davim, Eds. UK: Elsevier, 2020, pp. 13–39.

[7] A. Z. Gorski and J. Skrzat, "Error estimation of the fractal dimension measurements of cranial sutures," *J. Anat.*, vol. 208, no. 3, pp. 353–359, Mar. 2006.

[8] N. Lynnerup and J. C. B. Jacobsen, "Brief communication: Age and fractal dimensions of human sagittal and coronal sutures," *Am. J. Phys. Anthropol.*, vol. 121, no. 4, pp. 332–336, 2003.

[9] M. Bouda, J. S. Caplan, and J. E. Saiers, "Box-counting dimension revisited: Presenting an efficient method of minimizing quantization error and an assessment of the self-similarity of structural root systems," *Front. Plant Sci.*, vol. 7, no. Feb, pp. 1–15, Feb. 2016.

[10] C. Panigrahy, A. Seal, N. K. Mahato, and D. Bhattacharjee, "Differential box counting methods for estimating fractal dimension of gray-scale images: A survey," *Chaos Solitons Fractals*, vol. 126, pp. 178–202, Sep. 2019.

[11] C. Panigrahy, A. Garcia-Pedrero, A. Seal, D. Rodríguez-Esparragón, N. K. Mahato, and C. Gonzalo-Martín, "An approximated box height for Differential-Box-Counting method to estimate fractal dimensions of gray-scale images," *Entropy*, vol. 19, no. 10, 2017.

[12] Y. Liu et al., "An improved differential box-counting method to estimate fractal dimensions of gray-level images," *J. Vis. Commun. Image Represent.*, vol. 25, no. 5, pp. 1102–1111, Jul. 2014.

[13] C. Panigrahy, A. Seal, and N. K. Mahato, "Quantitative texture measurement of gray-scale images: Fractal dimension using an improved differential box counting method," *Measurement*, vol. 147, p. 106859, Dec. 2019.

[14] D. Hong, Z. Pan, and X. Wu, "Improved differential box counting with multi-scale and multi-direction: A new palmprint recognition method," *Optik (Stuttg).*, vol. 125, no. 15, pp. 4154–4160, 2014.

[15] J. Li, Q. Du, and C. Sun, "An improved box-counting method for image fractal dimension estimation," *Pattern Recognit.*, vol. 42, no. 11, pp. 2460–2469, 2009.

[16] C. Panigrahy, A. Seal, and N. K. Mahato, "Image texture surface analysis using an improved differential box counting based fractal dimension," *Powder Technol.*, vol. 364, pp. 276–299, Mar. 2020.

[17] G.-B. So, H.-R. So, and G.-G. Jin, "Enhancement of the Box-Counting Algorithm for fractal dimension estimation," *Pattern Recognit. Lett.*, vol. 98, pp. 53–58, Oct. 2017.

[18] J. Yan, Y. Sun, S. Cai, and X. Hu, "An improved box-counting method to estimate fractal dimension of images," *J. Appl. Anal. Comput.*, vol. 6, no. 4, pp. 1114–1125, 2016.

[19] J. Jiménez and J. Ruiz de Miras, "Fast box-counting algorithm on GPU," *Comput. Methods Programs Biomed.*, vol. 108, no. 3, pp. 1229–1242, Dec. 2012.

[20] D. Ristanović, B. D. Stefanović, and N. Puškaš, "Fractal analysis of dendrite morphology using modified box-counting method," *Neurosci. Res.*, vol. 84, pp. 64–67, Jul. 2014.

[21] Z. Feng and X. Sun, "Box-counting dimensions of fractal interpolation surfaces derived from fractal interpolation functions," *J. Math. Anal. Appl.*, vol. 412, no. 1, pp. 416–425, Apr. 2014.

[22] K. Foroutan-pour, P. Dutilleul, and D. Smith, "Advances in the implementation of the box-counting method of fractal dimension estimation," *Appl. Math. Comput.*, vol. 105, no. 2–3, pp. 195–210, Nov. 1999.

[23] J. Xu, Z. Jian, and X. Lian, "An application of box counting method for measuring phase fraction," *Meas. J. Int. Meas. Confed.*, vol. 100, pp. 297–300, 2017.

[24] A. Tripathy, A. Kumar, V. Srinivasan, K. H. Singh, and T. N. Singh, "Fractal analysis and spatial disposition of porosity in major indian gas shales using low-pressure nitrogen adsorption and advanced image segmentation," *J. Nat. Gas Sci. Eng.*, vol. 72, no. September, p. 103009, Dec. 2019.

[25] B. H. Lee and S. K. Lee, "Effects of specific surface area and porosity on cube counting fractal dimension, lacunarity, configurational entropy, and permeability of model porous networks: Random packing simulations and NMR micro-imaging study," *J. Hydrol.*, vol. 496, pp. 122–141, 2013.

[26] M. Tanaka, Y. Kimura, L. Chouanine, R. Kato, and J. Taguchi, "Fractal analysis of the three-dimensional fracture surfaces in materials by the box-counting method," *J. Mater. Sci. Lett.*, vol. 22, no. 18, pp. 1279–1281, 2003.

[27] J. Weiss, "Self-affinity of fracture surfaces and implications on a possible size effect on fracture energy," *Int. J. Fract.*, vol. 109, no. 4, pp. 365–381, 2001.

[28] D. Yin, B. Chen, W. Ye, J. Gou, and J. Fan, "Mechanical test and fractal analysis on anisotropic fracture of cortical bone," *Appl. Surf. Sci.*, vol. 357, pp. 2063–2068, Dec. 2015.

[29] F. M. Mwema, E. T. Akinlabi, and O. P. Oladijo, "Fractal analysis of thin films surfaces: A brief overview," in *Advances in Material Sciences and Engineering. Lecture Notes in Mechanical Engineering*, Mokhtar Awang, Seyed Sattar Emamian, and Farazila Yusof, Eds. Singapore: Springer, 2020, pp. 251–263.

[30] W. Ju and N. S. N. Lam, "An improved algorithm for computing local fractal dimension using the triangular prism method," *Comput. Geosci.*, vol. 35, no. 6, pp. 1224–1233, 2009.

[31] K. C. Clarke, "Computation of the fractal dimension of topographic surfaces using the triangular prism surface area method," *Comput. Geosci.*, vol. 12, no. 5, pp. 713–722, 1986.

[32] Y. Zhou, T. Fung, and Y. Leung, "Improved triangular prism methods for fractal analysis of remotely sensed images," *Comput. Geosci.*, vol. 90, pp. 64–77, May 2016.

[33] W. C. Chow, "Fractal (fractional) Brownian motion," *Wiley Interdiscip. Rev. Comput. Stat.*, vol. 3, no. 2, pp. 149–162, 2011.

[34] D. Delignières, "Correlation properties of (discrete) fractional gaussian noise and fractional brownian motion," *Math. Probl. Eng.*, vol. 2015, pp. 1–7, 2015.

[35] M. Li, "On the long-range dependence of fractional brownian motion," *Math. Probl. Eng.*, vol. 2013, no. 4, pp. 1–5, 2013.

[36] R. P. Yadav et al., "Effect of angle of deposition on the Fractal properties of ZnO thin film surface," *Appl. Surf. Sci.*, vol. 416, pp. 51–58, 2017.

[37] R. Lopes and N. Betrouni, "Fractal and multifractal analysis: A review," *Med. Image Anal.*, vol. 13, no. 4, pp. 634–649, Aug. 2009.

[38] F. M. Mwema, O. P. Oladijo, T. S. Sathiaraj, and E. T. Akinlabi, "Atomic force microscopy analysis of surface topography of pure thin aluminium films," *Mater. Res. Express*, vol. 5, no. 4, pp. 1–15, Apr. 2018.

[39] R. Gavrila, A. Dinescu, and D. Mardare, "A power spectral density study of thin films morphology based on AFM profiling," *Rom. J. Inf. Sci. Technol.*, vol. 10, no. 3, pp. 291–300, 2007.

[40] F. M. Mwema, O. P. Oladijo, and E. T. Akinlabi, "The use of power spectrum density for surface characterization of thin films," in *Photoenergy and Thin Film Materials*, X.-Y. Yang, Ed. Hoboken, NJ: John Wiley & Sons, Inc., 2019, pp. 379–411.

[41] F. M. Mwema, E. T. Akinlabi, and O. P. Oladijo, "Two-dimensional fast fourier transform analysis of surface microstructures of thin aluminium films prepared by Radio-Frequency (RF) magnetron sputtering," in *Advances in Materials Science and Engineering Lecture Notes in Mechanical Engineering*, Mokhtar Awang, Seyed Sattar Emamian, and Farazila Yusof, Eds. Singapore: Springer, 2019, pp. 239–249.

[42] F. M. Mwema, O. P. Oladijo, and E. T. Akinlabi, "The use of power spectrum density for surface characterization of thin films," in *Photoenergy and Thin Film Materials*, X.-Y. Yang, Ed. Hoboken, NJ: John Wiley & Sons, Inc., 2019, pp. 379–411.

[43] F. M. Mwema, E. T. Akinlabi, and O. P. Oladijo, "Correction of artifacts and optimization of atomic force microscopy imaging," in *Title: Design, Development, and Optimization of Bio-Mechatronic Engineering Products*, K. Kumar and J. Paulo Davim, Eds. USA: IGI Global, 2019, pp. 158–179.

[44] F. M. Mwema, E. T. Akinlabi, and O. P. Oladijo, *"Complementary investigation of SEM and AFM on the morphology of sputtered aluminum thin films,"* in *Proceedings of the Eighth International Conference on Advances in Civil, Structural and Mechanical Engineering - CSM 2019*, Birmingham City, UK. 2019, pp. 10–14.

[45] F. M. Mwema, E. T. Akinlabi, O. P. Oladijo, S. A. Akinlabi, and S. Hassan, "Effect of AFM scan size on the scaling law of sputtered aluminium thin films," in *Advances in Manufacturing Engineering, Lecture Notes in Mechanical Engineering*, Mokhtar Awang, Seyed Sattar Emamian, and Farazila Yusof, Eds. 2020, pp. 171–176.

[46] F. M. Mwema, E. T. Akinlabi, and O. P. Oladijo, *"Influence of sputtering power on surface topography, microstructure and mechanical properties of aluminum thin films,"* in *Proceedings of the Eighth International Conference on Advances in Civil, Structural and Mechanical Engineering - CSM 2019*, Birmingham City, UK. 2019, pp. 5–9.

[47] F. M. Mwema, E. T. Akinlabi, and O. P. Oladijo, *"Exploring the effect of rf power in sputtering of aluminum thin films-a microstructure analysis,"* *Proceedings of the International Conference on Industrial Engineering and Operations Management*, 2019, pp. 745–750.

[48] N. B. Nezafat, M. Ghoranneviss, S. M. Elahi, A. Shafiekhani, Z. Ghorannevis, and S. Solaymani, "Topographic characterization of canine teeth using atomic force microscopy images in nano-scale," *Int. Nano Lett.*, vol. 9, no. 4, pp. 311–315, Dec. 2019.

[49] Y. L. Kong, S. V. Muniandy, K. Sulaiman, and M. S. Fakir, "Random fractal surface analysis of disordered organic thin films," *Thin Solid Films*, vol. 623, pp. 147–156, 2017.

[50] T. D. B. Jacobs, T. Junge, and L. Pastewka, "Quantitative characterization of surface topography using spectral analysis," *Surf. Topogr. Metrol. Prop.*, vol. 5, no. 1, p. 013001, 2017.

[51] D. Raoufi, "Fractal analyses of ITO thin films: A study based on power spectral density," *Phys. B Condens. Matter*, vol. 405, no. 1, pp. 451–455, 2010.

[52] H. Salat, R. Murcio, and E. Arcaute, "Multifractal methodology," *Phys. A Stat. Mech. its Appl.*, vol. 473, pp. 467–487, v2017.

[53] R. P. Yadav, M. Kumar, A. K. Mittal, and A. C. Pandey, "Fractal and multifractal characteristics of swift heavy ion induced self-affine nanostructured BaF 2 thin film surfaces," *Chaos An Interdiscip. J. Nonlinear Sci.*, vol. 25, no. 8, p. 083115, Aug. 2015.

[54] G. Durán-Meza, J. López-García, and J. L. del Río-Correa, "The self-similarity properties and multifractal analysis of DNA sequences," *Appl. Math. Nonlinear Sci.*, vol. 4, no. 1, pp. 267–278, 2019.

[55] L. G. Souza França et al., "Fractal and multifractal properties of electrographic recordings of human brain activity: Toward its use as a signal feature for machine learning in clinical applications," *Front. Physiol.*, vol. 9, no. December, pp. 1–18, 2018.

[56] W. Drzewiecki, A. Wawrzaszek, M. Krupiński, S. Aleksandrowicz, and K. Bernat, "Applicability of multifractal features as global characteristics of Worldview-2 panchromatic satellite images," *Eur. J. Remote Sens.*, vol. 49, pp. 809–834, 2016.

[57] F. M. Mwema, E. T. Akinlabi, and O. P. Oladijo, "Fractal analysis of hillocks: A case of RF sputtered aluminum thin films," *Appl. Surf. Sci.*, vol. 489, pp. 614–623, Sep. 2019.

[58] C. Liu, X. L. Jiang, T. Liu, L. Zhao, W. X. Zhou, and W. K. Yuan, "Multifractal analysis of the fracture surfaces of foamed polypropylene/polyethylene blends," *Appl. Surf. Sci.*, vol. 255, no. 7, pp. 4239–4245, 2009.

[59] S. Blacher, F. Brouers, and G. Ananthakrishna, "Multifractal analysis, a method to investigate the morphology of materials," *Phys. A Stat. Mech. its Appl.*, vol. 185, no. 1–4, pp. 28–34, Jun. 1992.

[60] X. Sun, Z. Fu, and Z. Wu, "Multifractal analysis and scaling range of ZnO AFM images," *Phys. A Stat. Mech. its Appl.*, vol. 311, no. 3–4, pp. 327–338, Aug. 2002.

[61] A. Biswas, H. P. Cresswell, and C. Si, "Application of multifractal and joint multifractal analysis in examining soil spatial variation: A review," in *Fractal Analysis and Chaos in Geosciences*, Sid-Ali Ouadfeulvol Eds., 395, no. tourism, InTech, 2012, pp. 116–124.

[62] W. J. Xie, Z. Q. Jiang, G. F. Gu, X. Xiong, and W. X. Zhou, "Joint multifractal analysis based on the partition function approach: Analytical analysis, numerical simulation and empirical application," *New J. Phys.*, vol. 17, no. 10, 2015, pp. 1–13.

[63] S. Banerjee, Y. He, X. Guo, and B. C. Si, "Spatial relationships between leaf area index and topographic factors in a semiarid grassland: Joint multifractal analysis," *Aust. J. Crop Sci.*, vol. 5, no. 6, pp. 756–763, 2011.

[64] J. Wang, P. Shang, and W. Ge, "Multifractal cross-correlation analysis based on statistical moments," *Fractals*, vol. 20, no. 3–4, pp. 271–279, 2012.

[65] A. N. Kravchenko, D. G. Bullock, and C. W. Boast, "Joint multifractal analysis of crop yield and terrain slope," *Agron. J.*, vol. 92, no. 6, pp. 1279–1290, 2000.

[66] D. C. Lin, "Factorization of joint multifractality," *Phys. A Stat. Mech. its Appl.*, vol. 387, no. 14, pp. 3461–3470, 2008.

[67] L. Y. He and S. P. Chen, "Multifractal detrended cross-correlation analysis of agricultural futures markets," *Chaos, Solitons and Fractals*, vol. 44, no. 6, pp. 355–361, 2011.

[68] M. S. Movahed, G. R. Jafari, F. Ghasemi, S. Rahvar, and M. R. R. Tabar, "Multifractal detrended fluctuation analysis of sunspot time series," *J. Stat. Mech. Theory Exp.*, vol. 316, no. 2, pp. 87–114, 2006.

[69] H. Liu, X. Zhang, and X. Zhang, "Multiscale multifractal analysis on air traffic flow time series: A single airport departure flight case," *Phys. A Stat. Mech. its Appl.*, vol. 545, 2020, pp. 1–17.

[70] Z. Fayyaz, M. Bahadorian, J. Doostmohammadi, V. Davoodnia, S. Khodadadian, and R. Lashgari, "Multifractal detrended fluctuation analysis of continuous neural time series in primate visual cortex," *J. Neurosci. Methods*, vol. 312, no. November 2018, pp. 84–92, 2019.

[71] X. Zhang, L. Yang, and Y. Zhu, "Analysis of multifractal characterization of Bitcoin market based on multifractal detrended fluctuation analysis," *Phys. A Stat. Mech. its Appl.*, vol. 523, pp. 973–983, 2019.

[72] S. Fang, X. Lu, J. Li, and L. Qu, "Multifractal detrended cross-correlation analysis of carbon emission allowance and stock returns," *Phys. A Stat. Mech. its Appl.*, vol. 509, pp. 551–566, 2018.

[73] G.-F. Gu and W.-X. Zhou, "Detrended fluctuation analysis for fractals and multifractals in higher dimensions," *Phys. Rev. E*, vol. 74, no. 6, p. 061104, Dec. 2006.

[74] W. Wang, K. Liu, and Z. Qin, "Multifractal analysis on the return series of stock markets using MF-DFA method," *IFIP Adv. Inf. Commun. Technol.*, vol. 426, no. October 1987, 2014, pp. 107–115.

[75] E. Oral and G. Unal, "Modeling and forecasting time series of precious metals: A new approach to multifractal data," *Financ. Innov.*, vol. 5, no. 1, 2019.

[76] F. Agterberg, "Multifractal modeling of worldwide and Canadian metal size-frequency distributions," *Nat. Resour. Res.*, vol. 29, no. 1, pp. 539–550, 2020.

[77] L. Catalano and A. Figliola, "Analysis of the nonlinear relationship between commodity prices in the last two decades," *Qual. Quant.*, vol. 49, no. 4, pp. 1553–1558, 2015.

[78] Z. Li and X. Lu, "Multifractal analysis of China's agricultural commodity futures markets," *Energy Procedia*, vol. 5, pp. 1920–1926, 2011.

[79] S. P. Chen and L. Y. He, "Multifractal spectrum analysis of nonlinear dynamical mechanisms in China's agricultural futures markets," *Phys. A Stat. Mech. its Appl.*, vol. 389, no. 7, pp. 1434–1444, 2010.

[80] H. Kim, G. Oh, and S. Kim, "Multifractal analysis of the Korean agricultural market," *Phys. A Stat. Mech. its Appl.*, vol. 390, no. 23–24, pp. 4286–4292, Nov. 2011.

[81] L.-Y. He and S.-P. Chen, "Are developed and emerging agricultural futures markets multifractal? A comparative perspective," *Phys. A Stat. Mech. its Appl.*, vol. 389, no. 18, pp. 3828–3836, Sep. 2010.

[82] M. A. Gonçalves, "Characterization of geochemical distributions using multifractal models," *Math. Geol.*, vol. 33, no. 1, pp. 41–61, 2001.

[83] E. Muñoz, G. Poveda, A. Ochoa, and H. Caballero, *Advancing Culture of Living with Landslides*, Switzerland: Springer International Publishing. vol. 1984. 2017.

[84] S. Xie and Z. Bao, "Fractal and multifractal properties of geochemical fields," *Mathematical Geology*, vol. 36, no. 7. pp. 847–864, 2004.

[85] L. Tennekoon, M. C. Boufadel, and J. E. Nyquist, "Multifractal characterization of airborne geophysical data at the Oak Ridge facility," *Stoch. Environ. Res. Risk Assess.*, vol. 19, no. 3, pp. 227–239, 2005.

[86] H. F. Jelinek, N. T. Milošević, A. Karperien, and B. Krstonošić, "Box-counting and multifractal analysis in neuronal and glial classification," *Adv. Intell. Syst. Comput.*, vol. 187 AISC, pp. 177–189, 2013.

[87] B. J. West and N. Scafetta, "A multifractal dynamical model of human gait," in *Fractals in Biology and Medicine*, Gabriele A. Losa, Danilo Merlini, Theo F. Nonnenmacher, and Ewald R. Weibel, Eds., Basel, Switzerland: Birkhäuser Basel. 2006, pp. 131–140.

[88] P. D. Zegzhda, D. S. Lavrova, and A. A. Shtyrkina, "Multifractal analysis of internet backbone traffic for detecting denial of service attacks," *Autom. Control Comput. Sci.*, vol. 52, no. 8, pp. 936–944, 2018.

[89] M. Stehlík, P. Hermann, S. Giebel, and J.-P. Schenk, "Multifractal analysis on cancer risk," in *Recent Studies on Risk Analysis and Statistical Modeling*, Teresa A. Oliveira, Christos P. Kitsos, Amílcar Oliveira, and Luís Miguel Grilo, Eds, Springer International Publishing: Switzerland. 2018, pp. 17–33.

[90] H. Li, M. L. Giger, O. I. Olopade, and L. Lan, "Fractal analysis of mammographic parenchymal patterns in breast cancer risk assessment," *Acad. Radiol.*, vol. 14, no. 5, pp. 513–521, May 2007.

[91] T. Takahashi et al., "Multifractal analysis of deep white matter microstructural changes on MRI in relation to early-stage atherosclerosis," *Neuroimage*, vol. 32, no. 3, pp. 1158–1166, 2006.

[92] Y. Wei, X. Wu, J. Xia, and C. Cai, "Relationship between granitic soil particle-size distribution and shrinkage properties based on multifractal method," *Pedosphere*, vol. 30, no. 6, pp. 853–862, 2020.

[93] X. Ju, Y. Jia, T. Li, L. Gao, and M. Gan, "Morphology and multifractal characteristics of soil pores and their functional implication," *Catena*, vol. 196, no. December 2019, p. 104822, Jan. 2021.

[94] N. Bird, M. C. Díaz, A. Saa, and A. M. Tarquis, "Fractal and multifractal analysis of pore-scale images of soil," *J. Hydrol.*, vol. 322, no. 1–4, pp. 211–219, 2006.

[95] F. San José Martínez et al., "Multifractal analysis of discretized X-ray CT images for the characterization of soil macropore structures," *Geoderma*, vol. 156, no. 1–2, pp. 32–42, Apr. 2010.

[96] I. G. Torre, J. C. Losada, R. J. Heck, and A. M. Tarquis, "Multifractal analysis of 3D images of tillage soil," *Geoderma*, vol. 311, pp. 167–174, 2018.

[97] S. Verrier, "Multifractal and multiscale entropy scaling of in-situ soil moisture time series: Study of SMOSMANIA network data, southwestern France," *J. Hydrol.*, vol. 585, no. January, 2020.

[98] De Wang, B. Fu, W. Zhao, H. Hu, and Y. Wang, "Multifractal characteristics of soil particle size distribution under different land-use types on the Loess Plateau, China," *Catena*, vol. 72, no. 1, pp. 29–36, Jan. 2008.

[99] A. Biswas, "Joint multifractal analysis for three variables: Characterizing the effect of topography and soil texture on soil water storage," *Geoderma*, vol. 334, no. July 2018, pp. 15–23, Jan. 2019.

[100] B. B. Mandelbrot, "Self-affine fractal sets, I: The basic fractal dimensions," in *Fractals in Physics*, Luciano Pietronero and Erio Tosatti, Eds, UK: Elsevier, 1986, pp. 3–15.

[101] B. Mandelbrot, "How long is the coast of Britain? Statistical self-similarity and fractional dimension," *Science (80-.).*, vol. 156, no. 3775, pp. 636–638, May 1967.

[102] B. B. Mandelbrot, "Self-affine fractals and fractal dimension," *Phys. Scr.*, vol. 32, pp. 257–260, 1985.

[103] B. J. Florio, P. D. Fawell, and M. Small, "The use of the perimeter-area method to calculate the fractal dimension of aggregates," *Powder Technol.*, vol. 343, pp. 551–559, 2019.

[104] L. A. Brinkhoff, C. von Savigny, C. E. Randall, and J. P. Burrows, "The fractal perimeter dimension of noctilucent clouds: Sensitivity analysis of the area–perimeter method and results on the seasonal and hemispheric dependence of the fractal dimension," *J. Atmos. Solar-Terrestrial Phys.*, vol. 127, pp. 66–72, May 2015.

[105] A. R. Imre, "Artificial fractal dimension obtained by using perimeter–area relationship on digitalized images," *Appl. Math. Comput.*, vol. 173, no. 1, pp. 443–449, Feb. 2006.

[106] Z. Chen, Y. Liu, and P. Zhou, "A comparative study of fractal dimension calculation methods for rough surface profiles," *Chaos, Solitons and Fractals*, vol. 112, pp. 24–30, 2018.

[107] J. Li and F. Nekka, "Is the classical autocorrelation function appropriate for spatial signals defined on fractal supports?," *Phys. A Stat. Mech. its Appl.*, vol. 376, no. 1–2, pp. 147–157, Mar. 2007.

[108] R. C. Bokun, "Evaluation of the density autocorrelation function for fractal aggregates from the light scattering data," *J. Aerosol Sci.*, vol. 26, pp. S933–S934, Sep. 1995.

[109] F. M. Mwema, E. T. Akinlabi, and O. P. Oladijo, "Effect of substrate type on the fractal characteristics of AFM images of sputtered aluminium thin films," *Mater. Sci.*, vol. 26, no. 1, pp. 49–57, Nov. 2019.

[110] A. Grayeli Korpi et al., "Minkowski functional characterization and fractal analysis of surfaces of titanium nitride films," *Mater. Res. Express*, vol. 6, no. 8, p. 086463, Jun. 2019.

[111] I. Levchenko, J. Fang, K. (Ken) Ostrikov, L. Lorello, and M. Keidar, "Morphological characterization of graphene flake networks using minkowski functionals," *Graphene*, vol. 05, no. 01, pp. 25–34, 2016.

[112] N. Spyropoulos-Antonakakis et al., "Selective aggregation of PAMAM dendrimer nano-carriers and PAMAM/ZnPc nanodrugs on human atheromatous carotid tissues: A pho-todynamic therapy for atherosclerosis," *Nanoscale Res. Lett.*, vol. 10, no. 1, pp. 1–19, 2015.

[113] D. Legland, K. Kiêu, and M.-F. Devaux, "Computation of minkowski measures on 2D and 3D binary images," *Image Anal. Stereol.*, vol. 26, no. 2, p. 83, 2011.

[114] M. Salerno and M. Banzato, "Minkowski measures for image analysis in scanning probe microscopy," *Microsc. Anal.*, vol. 19, no. 4, pp. 13–15, 2005.

[115] X. Li, P. R. S. Mendonça, and R. Bhotika, "Texture analysis using Minkowski functionals," in *SPIE 8314, Medical Imaging 2012: Image Processing; 83144Y*, San Diego, California, USA. 2012, p. 83144Y.

Part 2

---

*Typical Studies of Fractal
Descriptions of Sputtered Films*

4 Fractal Characterization of Hillocks and Porosity in Sputtered Films

4.1 INTRODUCTION

Similar to most manufacturing technologies, thin film sputtering processes are characterized by structural defects, which affect their quality and performance. In Chapter 1, it was discussed that thin films sputtering is influenced by the following broad factors:

  i. The plasma processes,
 ii. The sputtering technology equipment factors, and
iii. The materials properties

It was stated that the proper combination of the sputtering parameters enhances the production of defect-free thin films. It can therefore be inferred that defects such as hillocks, porosity, and dislocations depending on the choice of the above sputtering parameters. In general, there are several defects in thin film deposition technologies and some of them are listed as follows:

    i. Hillocks
   ii. Porosity and voids
  iii. Random roughness of the boundaries
   iv. Inhomogeneity in properties
    v. Non-uniform thicknesses
   vi. Dislocations
  vii. Lattice defects
 viii. Incomplete coverage of the substrate

There are several challenges associated with thin film deposition which can contribute considerably to the formation of the above defects in thin film surfaces. These challenges have been discussed in detail by Şimşek et al. [1] in their work titled, 'Difficulties in Thin Film Synthesis' and some of them include:

  i. The characteristics of the materials used in the deposition process pose various challenges. Firstly, in physical vapor deposition (PVD) methods, the binding energy of the target material determines the energy required by the system to dislodge the material from that target. It may lead to a lack of thin film formation if the energy of the plasma is not enough to dislodge the target material. In the chemical vapor deposition (CVD) process, some materials such as rare earth metals (e.g. indium, neodymium, yttrium, erbium) are toxic and pose environmental challenges. Another important challenge that may lead to the formation of poor-quality thin films is the lack of proper adhesion of the target material onto the substrate surface. For example, depositing a thin film of materials with lattice mismatch with the substrate is likely to result in poor adhesion. For example, depositing ceramic AlN thin films on metallic or polymeric substrates is limited by the poor adhesion.

ii. The surface properties of the substrate are very important on the quality of the deposited thin films. The surface of the substrates should be prepared to remove any dirt, oxides, and irregularities, which may interfere with the atom diffusion of the target material onto the substrate. Additionally, the correct choice of the substrate material should be undertaken, for example, silicon/ SiO_2 substrates have been shown to cause hillocks on metallic thin films. Similarly, depositing metallic films on glass substrates is limited by the formation of hillock structures. There may also occur reactions between the substrate and thin film material and the formation of unnecessary structures on the thin film. In other cases, the conditions of deposition may not favor the use of some substrate materials. For instance, although polymeric materials are very attractive for the manufacturing of flexible electronic conductors, they are unsuitable as substrates for high temperature thin film deposition conditions.

iii. The challenges associated with thin film deposition devices/techniques can also lead to the creation of defects on thin films. For instance, during sputtering, there may occur contamination of the vacuum chamber, substrate, and target holders. This is due to the high energy required, the nature of the process, and long deposition time. The contamination is inevitable in most devices (especially in PVD systems) and it indeed presents a challenge. It is always advisable to clean the vacuum chamber, sample holders, and other parts of the system after every deposition and before the next thin film production to avoid contamination of the deposited thin film. It is not possible to remove all these contaminants in the deposition chamber and the chances, are when the process is running, the contaminants may be activated to affect the quality of the thin films. In CVD methods, the depositions are usually undertaken at high temperatures and this may lead to damage to the substrate and influence, unnecessarily, the formation of the thin film structures.

iv. There is an increasing need for hybrid, composite, and multilayer thin films, which requires the deposition of the various materials at the same time. In PVD, different materials exhibit different deposition characteristics, whereas, in CVD, materials would react differently when exposed to similar conditions. These may lead to challenges of adhesion between adjacent layers and defects such as interlayer porosity.

There are several studies that have reported on defects in thin films prepared by various techniques [2–5]. For instance, an early study by Qian, Skowroski, and Greg (1996) reported on the structural defects of GaN thin films grown through a chemical process [6]. It was shown that GaN thin films exhibited dislocations, which were attributed to the strains caused by the material mismatch between the films and the substrates. The defects were shown to depend on the thickness of the grown thin films; at a very small thickness, the dislocation defects exhibited irregular orientation whereas, at a larger thickness of the films, the dislocations were oriented in the [0001] planar direction (Figure 4.1a). Other defects likely to occur on the interfaces of the GaN/substrate are micro-twins and double-positioning boundaries. The planar view using TEM of the GaN/Al_2O_3 surfaces (shown in Figure 4.1b) revealed threading dislocations (shown by white arrows) and nanopipes (shown by black arrow). These nanopipes appear as open cores of screw dislocations. In the same study, it was discussed that high density of threading dislocations were associated with the low angle grain boundaries (LAGBs) whereas low densities were associated with poorly formed grain structures (Figure 4.2).

The reduction of defects in thin films is the main objective of any deposition process since defects influence, negatively, the thin film properties such as electrical, optical, and strength [7]. A study by Nemoz et al. (2017) investigated the reduction

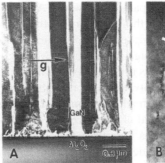

FIGURE 4.1 Dislocations in thin films. TEM micrographs of GaN films grown on Al_2O_3 substrates: (a) Cross-sectional view and (b) plan view (reused with permission from Cambridge University Press [6]).

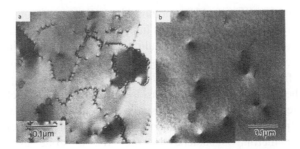

FIGURE 4.2 TEM micrographs on the surface of GaN/Sapphire thin films: (a) High density of threading dislocations and (b) low density of threading dislocations (reused with permission from Cambridge University Press [6]).

of dislocations in AlN thin films through annealing [8]. The AlN films were grown on sapphire substrates through molecular beam epitaxy and then annealed at a temperature range of 1350–1650°C. It was shown that annealing of AlN thin films within this range reduced the edge dislocation densities in their structure. Kappers et al. (2007) have illustrated the reduction of threading dislocations in (0001) GaN thin films by using SiN_x interlayers [9]. It was illustrated that the increase in the thickness of the interlayers led to a reduction in threading dislocations in GaN thin films. It was also demonstrated by Maeshashi et al. (1998) [10] that using thin Si substrates for deposition of GaAs films reduced the threading dislocation density. There are several other publications relating to threading dislocations and readers are referred to such literature [11–14].

The behavior, formation, and evolution of defects in thin films depend on the deposition parameters and conditions of the thin films. Although there exist a lot of publications on the quality of thin film deposition processes, there are few reports singly discussing the influence of deposition parameters on defects. The reason for scarce information in the subject could be due to the technological challenges associated with thin film characterizations. For instance, to study dislocations in thin films, at least a transmission electron microscopy (TEM) is required, which is not readily available to most researchers in different parts of the world. Additionally, preparatory equipment and procedures for TEM for thin films are costly and require expertise, which may not also be available to most researchers. As such, hillocks and porosity are the most studied in literature since high resolution SEM microscopy can detect these defects. In this chapter, the emphasis is therefore on hillocks and porosity as the common defects in thin films. Additionally, the authors of this book have worked on the fractal characterization of hillocks on Al thin films and rely on the expertise of such results. Studies related to the fractal characterization of hillocks and porosity are presented and important conclusions regarding the fractal characterization of these defects presented. The aim is to understand the growth mechanism of hillocks and porosity in relation to deposition conditions through fractal theory. The characterizations can also be used to detect the existence of the defects on the microstructure of the thin films.

4.2 HILLOCKS IN SPUTTERED THIN FILMS

Thin films grown by different PVD methods including sputtering exhibit defects, hillocks, and porosity are the main defects in these thin films. These defects interfere with the quality and performance/functionality of the thin films. For instance, nodular and trough defects have been shown to degrade the capability of thin films to protect the substrates from corrosion [15]. Hillocks usually occur as splats or surface extrusions on the microstructure of the films. In some literature, hillocks are described as 'surface blisters' on thin films, and in a study, the authors of this book revealed that these blisters 'pop-out' upon nanoindentation loading [16]. Their formation is attributed to the thermal mismatches between the substrate and film materials; mostly, hillocks occur when metallic films are deposited on non-metallic substrates such as glass and polymers. The structures usually occur along the grain boundaries and their formation is greatly dependent on grain size and crystal orientation. Since hillocks pose detrimental characteristics to the functionality of thin films such as poor electrical properties, there has been extensive research into ways of reducing them. Some of the common methods of reducing hillock formation in thin films are listed below.

i. Heat treatments: Several studies have shown that heat treatment can be used to minimize the hillock formations. However, the temperature of heat treatment should be carefully controlled since increasing the temperature continuously may lead to the growth of the size of the hillocks [17].

ii. Passivation layering: The use of capping materials on thin films reduces the hillock formation. For instance, the formation of hillocks in Al thin films grown on glass can be minimized via capping materials such as Ti, Mo, and SiO_2 [15]. However, the deposition of these passivation/capping layers on the thin films is not a straightforward affair as the thickness of the capping significantly influences their effectiveness. As such, optimization of the passivation process and layers is necessary for the effective reduction of hillocks.

iii. Dopants: The addition of %wt. of specific elements into the structure of the films can reduce the formation of hillock structures. For example, it was reported by Arai et al. [18] that addition of 2 wt.% Nd to Al films suppresses the formation hillocks. Additionally, it has been reported that doping Al films with copper lowers electromigration and therefore the formation of hillocks. The addition of transition metals such as Fe, Co, and Ni into the thin films of Al lowers the hillock formation in those films [15]. The addition of minute quantities of Scandium has also been shown to reduce the formation of hillocks although with a reduction in electrical conductivity [19]. It, therefore, implies that doping elements should be carefully selected to avoid significant degradation of the properties of thin films.

There are various studies describing the formation of hillocks in thin films prepared via sputtering technologies. One such study is the one by Lee et al. (2011), in which the suppression of hillock formation on Al thin films at varying quantities of scandium elements were reported [19]. It was reported that increasing the scandium quantities in Al thin films reduced the density of hillocks. Another study by Nam, Choi, and Lee (2000) [20] investigated the formation of hillocks in sputtered PZT films and indicated that the use of Ti underlayer resulted in an increase in hillock density. The effect of substrate temperature on hillock density and morphology for Al thin films was investigated by the authors of this book [16] and the results shall be discussed in detail in the Section 4.3. Ma et al. (2015) reported on the growth of hillocks in CuZr metallic glass materials [21]. Resnik and others (2012) investigated the influence of target composition and thermal treatment of Al alloy thin films sputtered on silicon wafers [22] and revealed that the composition of the target significantly influences the formation of hillocks during sputtering. Nazarpour et al. (2009) [23] investigated the influence of quenching in different media on the evolution of hillocks and electrical properties of Au/Pd prepared via electron beam deposition. In a related study,

Nazarpour (2012) reported on the morphological and fractal characteristics of hillocks in palladium thin films [24]. The effect of SiO_2 passivation layer thickness on hillock evolution was reported by Kim [25] and it was shown that 250 nm was enough to prevent the growth of hillocks on DC magnetron sputtered Al thin films. In a related study, Hwang et al. [26] investigated the influence of film thickness and annealing temperature on the formation of hillocks in Al thin films. Through a statistical analysis of the sizes of the hillocks, it was reported that an increase in the film's thickness results in a decrease in density and an increase in the size of the hillocks in pure Al thin films. Iwamura, Ohnishi, and Yoshikawa (1995) investigated the evolution of hillocks (with varying annealing temperature) on Al-Ta alloy films deposited via DC magnetron sputtering [27]. The in-situ microscopic characterization of the hillock structures revealed that the size of the hillocks of Al-Ta alloy films increased with the annealing temperature of the films. It was also shown that the addition of Ta into the Al-Ta alloy refined the grains of the film structure and reduced the number of hillocks. In another detailed study by Martin et al. [28], a comparative investigation of hillock formation on sputtered Al alloy films of different compositions (pure Al, Al-1.5 wt.% Cu, AlCu with 0.2 wt.% W, and AlCu with 0.4 wt.%). The temperature of sputtering was shown to be a very important parameter influencing the density of hillocks. It was also reported that addition of W increases the density of hillocks in AlCu alloy thin films whereas addition of Cu in the alloy films reduces the density of hillocks.

In the next section, studies related to the fractal characterization of hillock defects in thin films grown via sputtering technology have been discussed.

4.3 FRACTAL CHARACTERIZATION OF HILLOCKS IN SPUTTERED METALLIC/ALLOY THIN FILMS

In this section, the application of fractal theory in studying the character of hillocks in thin films is illustrated. The authors of this book have published an article titled, 'Fractal analysis of hillocks: A case of radio-frequency (RF) sputtered aluminium thin films' in the journal of *Applied Surface Science* [16]. The article approach is used here to illustrate the use of fractal theory in the characterization of hillocks in sputtered thin films. Both the mono-fractal and multifractal approaches can be suitably used to characterize the structure of hillocks on thin films as illustrated in the article.

In brief, the aluminium thin films in the reference article [16] were deposited via RF magnetron sputtering at three different substrate temperatures (55°C, 65°C, and 95°C) and constant time and power. The films were grown on glass substrates and evaluated for microstructural, topography, and mechanical characteristics. The preparation of the substrates and deposition process was carefully undertaken before the deposition to avoid any challenges which may result to defects as earlier discussed. It has been reported in the literature that hillocks can occur during the deposition of the films, especially due to the mismatch between the substrate and the coating materials, post-deposition heat treatment (such as annealing), and during applications [15]. The hillocks discussed in the

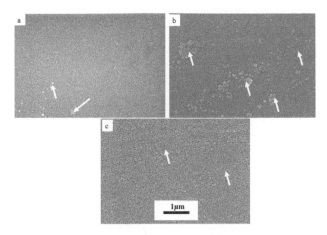

FIGURE 4.3 FESEM micrographs of Al thin films deposited on glass at different substrate temperature: (a) 55°C, (b) 65°C, and (c) 95°C (reused with permission from Elsevier Ltd).

article by Mwema et al. [16] occurred during the sputtering of Al films on glass substrates. On deposition, the films were imaged using AFM and FESEM and the FESEM micrographs were utilized in the fractal characterization of the hillocks. The formulations of box counting methodology and multifractal theory discussed in Chapter 3 were implemented in the FracLac tool in Fiji software [29,30].

The FESEM micrographs of the films identifying the hillock structures (white arrows) are shown in Figure 4.3. The hillock structures appear as blisters on the surface of the films. As shown, the hillocks are seen as small and isolated spherical particles at 55°C, whereas at 65°C, the hillocks appear clustered and enlarged. The density of the hillocks is seen to increase with the sputtering substrate temperature. At the highest temperature, the hillocks appear well organized and distributed as round particles within the pure Al structure.

The mono-fractal characterizations of the hillock structures were undertaken using an area (A)–perimeter (P) method. In this method, the images were segmented, through the thresholding procedure, and then using ImageJ (now Fiji) software, particle analysis was undertaken from which area fraction and perimeters were obtained. A large population of the particles was used for statistical accuracy. Plots of perimeter versus area were generated and then, using a power law relationship, $P = kA^{(D-1)}$, where D is the fractal dimension, and k is a constant of proportionality, the fractal dimensions of the three samples were computed. The fractal dimensions of the hillocks were shown to increase with an increase in the substrate temperature of the sputtering process.

The multifractal evaluation of the hillock structures was undertaken using mass exponent, singularity spectrum, and generalized multifractal dimension functions (described in Chapter 3). The results of multifractal characterizations for the hillocks as reported in the paper by the authors of this book [16] are represented by Figures 4.4, 4.5, and 4.6 for singularity spectrum, generalized fractal dimension, and mass

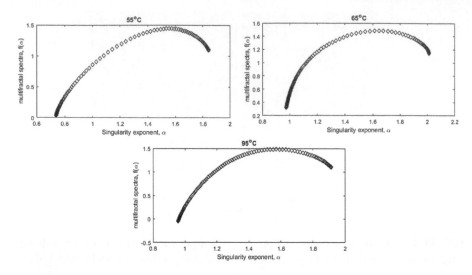

FIGURE 4.4 Multifractal spectrum as a function of singularity exponent for hillock features on Al thin films grown on glass substrates at different substrate temperatures (reused with permission from Elsevier) [16].

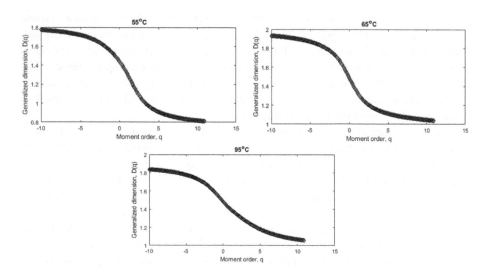

FIGURE 4.5 Generalized multifractal dimension against the moment order for hillock structures formed on Al thin films at different temperatures of the substrate (reused with permission from Elsevier) [16].

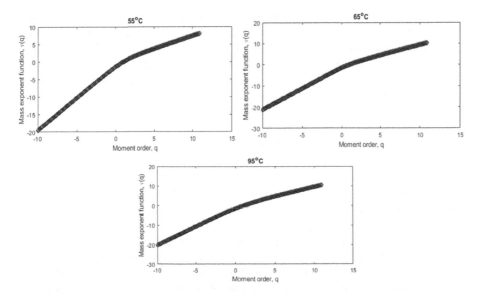

FIGURE 4.6 Mass exponent function against the moment order for the hillock structures on Al thin films sputtered on glass substrates at three different temperatures (reused with permission from Elsevier) [16].

moment exponent functions, respectively. From these results, the following conclusions can be drawn regarding fractal characterizations of hillock structures in thin film surfaces:

i. The multifractal scaling exponent $\tau(q)$ exhibited a nonlinear increase with the moment order, q, indicating the multifractality of the hillock structures on Al thin films. The nature of the linearity of the mass exponent functions indicates the interconnectivity of the hillock structures.

ii. The largest widths of the multifractal spectrum, which is an indicator of the strength of multifractality of features, occurs on the hillock structures with small sizes and density. The existence of smaller sizes of hillocks on the matrix of the main microstructure causes higher roughness and lateral deviation of the surface features. High values of Δf indicate the presence of well-developed, smooth, and complex hillock structures. For instance, in the article [16], Mwema et al. reported that at 95°C, the Al thin films consisted of smooth and complex structures of hillocks whereas at 65°C, the hillocks were characterized by rough and well-developed hillock structures as earlier seen in Figure 4.3. In Chapter 5, the relationship between the surface roughness and fractal characteristics of thin film features have been discussed and can be related to this description.

iii. The parameters from the generalized multifractal dimension for characterization of features are capacity dimension (D, at q = 0), the fractal dimension of the hillocks (D, q = 1), and correlation dimension (D, q = 2). The
difference between the capacity dimension and fractal dimension indicates
the heterogeneity in distribution and large values of the difference are indicators of non-uniform distribution of large hillock structures on the film
structure.

Nazarpour and Chaker (2012) employed the fractal theory to characterize hillocks
in palladium thin films prepared at different film thicknesses [24]. The topography of
the Pd thin films were taken using AFM in tapping mode. The AFM images were
then taken through surface flooding to isolate the background and expose the hillocks
on the surface of the films. Then, the perimeter (P) and surface area (S) of each
hillock were determined. Then, the area–perimeter method of computing fractal
dimension was used in which the plots of perimeter against the surface area were
generated on a bi-logarithmic scale. The points on such plots are usually representative
of the dimensionality of the hillocks and the distribution of their sizes within the
structure of the thin films. Then, a power law relating the perimeter and surface area,
$P = kS^{\frac{D-1}{2}}$, was applied to the plotted data to determine the fractal dimension (D) of
the hillocks. The fractal dimensions of the hillocks were shown to increase with the
thickness of the films although it started to reduce beyond some thickness. The
average height of the hillocks was shown to increase, similarly to the fractal
dimensions with the film thickness.

Another interesting feature of fractal characterization is the detection of defects/
abnormalities on thin film surfaces. The concept of detecting defects on thin films is
based on the 'disruption of the structural patterning' as a result of the introduction
of defects into the microstructure. For instance, a study by Shidpour and Movahed
[31] reported on the use of multifractal analysis in identifying defects in 2D semiconductors using MoS_2 monolayer. In most cases, photoluminescence has been used
to characterize single layer thin films; however, it is limited in that it cannot differentiate between defective and non-defective thin films. As such, this study [31] car-
ried out multifractal detrended fluctuation analysis (MF-DFA) and multifractal
detrended moving average analysis (MF-DMA) on the photoluminescence spectrum
for both defective and non-defective samples. There were four samples of thin films
of MoS_2, one which was non-defective and the other three were defective. The
defects on the films were created by ion sputtering of their surfaces at 1 cycle, 2
cycle, and 3 cycles. To undertake the multifractal analysis of the films, MF-DFA and
MF-DMA were applied on the incremental data obtained from the photoluminescence spectrum for all the samples. The following important points can be deduced
from this approach:

i. The plots of fluctuation functions (F_q) against the number of windows (s) were
shown to exhibit linear behavior. All the plots were accurately fitted to a power-
law function. This means that the photoluminescence data were multifractal
and the methodology could be used for the analyses of the thin films.

ii. The F_q plots for both methods (MF-DFA and MF-DMA) at the same values of q were close and therefore MF-DMA can be adopted (solely) for the multifractal analysis of the thin films.

iii. The slope of the F_q against s was determined to be the highest on the non-defective samples and generally decreased with the increasing number of cycles for ion sputtering. It, therefore, means samples with lower values of the slope (F_q vs. s) could be an indication of defects.

iv. The plots of generalized Hurst exponent (h(q)) against q exhibited a decreasing relationship of the function with increasing q. This indicates that the photoluminescence (PL) data for all the samples were multifractal and hence the samples surfaces.

v. The plots of the generalized Hurst exponents h(q) were all overlapping at q<0 for all the samples but there were clear differences at q>0. The values of the Hurst exponents for all the samples, at q = 2, were greater than 1. This means that the PL data were non-stationary. In such cases, the Hurst exponent can be computed as H = h(q = 2)−1.

vi. The values of the H computed at q = 2, were all greater than 0.5, which indicates that the PL data sets were classified in long-range correlated process. Generally, the Hurst exponent values decrease with the increase in the number of defects.

vii. As the defects increases in the structure of the films, the H (q = 2) approaches to 0.5 indicating a tendency to uncorrelated behavior.

viii. The slope of F_q (at q > 0) for the PL data reduces as the number of defects increased. The fluctuations are attributed to the coupling of phonons which is more pronounced in atoms close to the defects. In MoS_2, the defects usually occur as point defects (vacancy) as a result of the missing sulfur in the structure.

ix. The plots on singularity spectrum (f(α)) against the Holder exponent (α) for the non-defective and defective samples were also generated and it was shown that all the samples exhibited a left-hook profile. As the defective population increases, the plot tends toward symmetry.

x. As the number of defects increases, the width of the singularity spectra increases, which means that there is a broadening of the spectra, and hence stronger multifractality. The larger the number of defects in the films, the stronger the multifractal characteristics. The defects create a structural disruption and re-ordering of the patterning of the features. As such, the large number of defects makes the microstructure more complex and therefore leading to a stronger multifractal character.

xi. The position of the f_{max}(α) determines the extent of fluctuations of the surface features; as the number of defects increases, the position shifts to the right. This means that at a higher number of defects, the singularity spectrum is dominated by finer fluctuations.

xii. The plots of multifractal scaling exponent τ(q) = qh(q)−1 against q revealed that as the number of defects increases (at q > 0), the slope of the function decreases considerably.

xiii. The plots of generalized multifractal dimensions for the non-defective and defective films indicate a monotonic decrease of D(q) with q for all the PL data. However, at different values of q, the values of D(q) decrease with the increasing number of point defects in the structure of the films.

4.4 FRACTAL STUDIES OF POROSITY IN SPUTTERED METALLIC/ ALLOY THIN FILMS

Porosity in structures of thin films can emerge as a design objective of the films or due to limitations of the processing technology. As a design objective, the films are intentionally produced with porous structures, whereas in the latter scenario, the porosity is treated as a defect. Regardless of the case, studying these properties in thin films is crucial for the quality control and performance of the films. There are several studies discussing generally porosity in thin films [32–45]. The investigation of the porosity is important for specific films to comprehend their formation and mitigation where necessary. In this course, the porosity of the films should be imaged or measured for investigation and analyses. Most of the common existing techniques of analyzing/measuring porosity include microscopy, radiation scattering, gas adsorption, X-ray porosimeter, positron annihilation lifetime spectroscopy, quartz crystal microbalance, and ellipsometry [46,47] The X-ray porosimeter, quartz crystal microbalance, ellipsometry porosimeter, etc., have been developed specifically for thin film analyses due to small analyte of the thin films. The porosity of thin films depends on the structure and deposition conditions of the films and therefore their evaluation is very important.

A study by Oudrhiri-Hassani et al. (2008) investigated the structure, porosity, and roughness of RF magnetron sputtered oxide thin films and revealed that the sputtering pressure and target–substrate distance influence porosity of sputtered oxide films [45]. In this study, ellipsometry was used in porosity measurement of the films and revealed that films deposited at an argon pressure of 0.5 Pa and 5 cm target–substrate distance were dense and did not have any porosity. A similar study by Kumar et al. (2020) reported the dependence of porosity of sputtered $Si_{1-x}Sn_x$ thin films on argon gas pressure [48]. It was reported, however on the contrary, that films deposited at sputtering pressure and larger separation of target and substrate exhibited porous structures. As per the study [45], and the literature published by the authors of this book [15], the structure of sputtered films can be idealized into three layers as shown in Figure 4.7. As shown, the layer next to the substrate consists of a fully dense

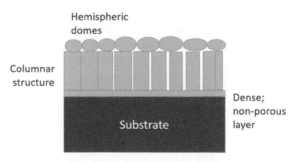

FIGURE 4.7   Model of the microstructure of sputtered thin films [45].

structure without porosity. The next layer consists of columnar structures, which form a compact plane. The layer has a porosity which can be attributed to lattice interstices [45]. Finally, the top layer consists of hemispherical domes covering the columnar structures of the second layer. These domes are characterized by the high surface roughness of the sputtered thin films. The top and second layers are characteristic of the Thornton's SZM (discussed in Chapter 1) when the sputtering occurs at low substrate temperature. As per the model, the formation of various layers and porosity will depend on the conditions of deposition and quality of the substrate and target materials. The type of porosity, either open or closed, will also depend on these conditions. In most of the existing studies, statistical analyses from imaging/microscopy are used such that it is possible to relate the surface roughness to the porosity evolution. Fractal characterization of porosity could be used to understand the lateral evolution of such structures and their relationship with surface roughness. In this section, the fractal characterizations of porosity in sputtered thin films are presented and important conclusions are drawn.

A study by Chen and Kitai (2008) [49] reported on a novel method of producing porous SiO_2 through the sputtering technique. The films were deposited on the porous substrate at a temperature of 200°C. It was discussed that the porosity of the films depends on the porosity of the porous substrate and the thickness of the film. In the study, films with four different thicknesses were deposited: 80 nm, 160 nm, 250 nm, and 350 nm (Figure 4.8). The films at 80 nm thickness were discontinuous and had preferential nucleation on top of the porous substrate. As seen in Figure 4.8, the size of the porosity decreased gradually with the increasing film thickness. The study by Chen and Kitai [49] only undertook morphological descriptions of the porous structures. In this section, their microscopy data were used to illustrate the fractal description of porosity for thin film structures of SiO_2.

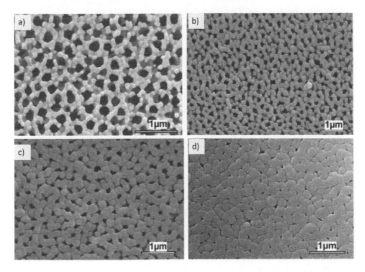

FIGURE 4.8 SEM images of thin films of SiO_2 sputtered on porous alumina at different thicknesses: (a) 80 nm, (b) 160 nm, (c) 250 nm, and (d) 350 nm (reused with permission from Elsevier Limited [49]).

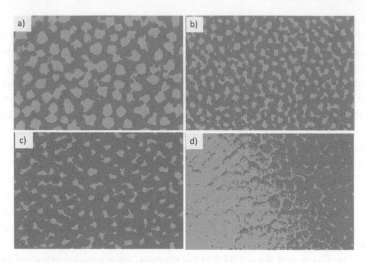

FIGURE 4.9 Binarized images of the SiO$_2$ thin films deposited on porous alumina at different thicknesses: (a) 80 nm, (b) 160 nm, (c) 250 nm, and (d) 350 nm.

Firstly, the obtained images were segmented through the thresholding technique, and the resulting images are shown in Figure 4.9. As shown, the porous features were analyzed (foreground) while the structure of the substrate was isolated in the background. The multifractal theory (described in Chapter 3) was applied and implemented in FracLac plugin in Fiji software. The methodology used is similar to that used in the article described in Section 4.3 on hillocks [16].

Secondly, the fractal dimensions of the binary images (Figure 4.9) were determined through the box counting method (box counting method was detailed in Chapter 3). The fractal dimensions obtained have been summarized in Table 4.1. The value of fractal dimension (D) quantifies, singly, the length of a complex structure, such as the porosity in this case. It can be seen that the values increased with the increasing film thicknesses. This is because as the film thicknesses increased, the porosity became finer, interconnected, and more complex as can be visually observed in Figure 4.8. However, fractal measurements at different scales of the images presented here could not be undertaken to verify the mono-fractality of the

TABLE 4.1
Fractal Dimensions of the SiO$_2$ thin Films Sputtered on Porous Alumina at Different Film's Thicknesses

No.	Film Thickness (nm)	Fractal Dimension (D)
a)	80	1.71±0.02
b)	160	1.73±0.02
c)	250	1.72±0.02
d)	350	1.76±0.02

structures, and as such the binary images were further subjected to multifractal measurements for detailed analysis.

Next, the binary images were characterized for multifractality using the functions described in Chapter 3 (namely, the $f(\alpha)$ vs. α, $\tau(q)$ vs. q, $D(q)$ vs. q, and $h(q)$ vs. q). Figure 4.10 shows the generalized multifractal dimension, $D(q)$, as a function of the moment order (q) for the porous structures in SiO_2 thin films deposited at different thicknesses. As can be seen, the functions exhibit a monotonic decrease with increasing moment order ($-10<q<10$). These plots indicate the existence of the multifractality character of the porosity structures in the films. Figure 4.11 shows the singularity spectrum function plots for the samples. It can be seen that all the plots exhibited a hook-like shape with a hump and skewed to the right. These results confirm that the structures discussed here exhibit multifractal behavior. The multifractal characteristics of the films were obtained from the plots to measure the multifractal strength. First, the widths of the singularity spectrum, denoted as $\Delta\alpha$, were determined as 0.85, 0.76, 0.95, and 0.89 for samples of thicknesses 80 nm, 160 nm, 250 nm, and 350 nm, respectively. As it will be further detailed in Chapter 6, the parameter $\Delta\alpha$ is a measure of multifractality. It can be seen that generally, as the thickness of the films increased and the porosity size decreased, the width of the multifractal spectrum increased. The films with the lowest density and sizes of the porosity (250 and 350 nm) exhibit the largest values of $\Delta\alpha$, indicating stronger multifractality. The results indicate that in suspiciously porous structures, the singularity spectrum function can be used to classify the size and density of the porosity in the films. The other parameters of importance in the spectrum are $f_{max}(\alpha)$ and Δf. According to Figure 4.11, Δf values were determined as 0.94, 0.65, 0.48, and 0.79 for the respective thicknesses of the SiO_2 films. The Δf parameter is a measure of the fractal dimension of the structure. It can be seen that the values were highest at the conditions of very high density and size of porosity (thickness of 80 nm) and lowest at the low density of porosity.

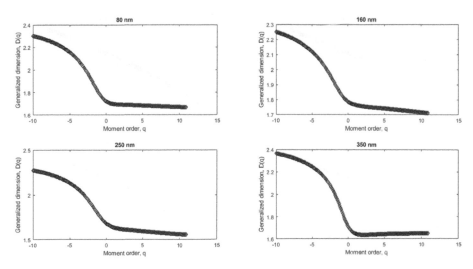

FIGURE 4.10   Generalized multifractal dimension D(q) for the four samples of SiO_2 thin films of different thicknesses.

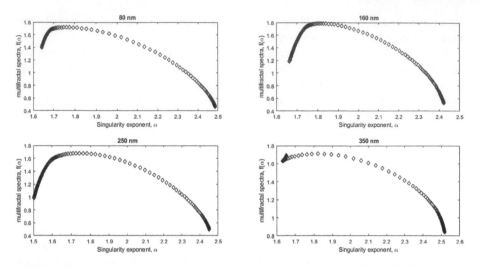

FIGURE 4.11 Singularity spectrum f(α) as a function of singularity exponent (α) for the SiO₂ thin films of different thicknesses.

The values are in agreement with the generalized fractal dimension, D(q), functions shown in Figure 4.10. The values of $f_{max}(\alpha)$ from the multifractal spectrum (Figure 4.11) were computed as 1.72, 1.83, 1.68, and 1.71, respectively, for 80 nm, 160 nm, 250 nm, and 350 nm thickness samples. The position of $f_{max}(\alpha)$ determines the nature of surface structures. In this case, positions of the $f_{max}(\alpha)$ are such that the profiles are tilted to the right indicating dominance by finer fluctuations. As the porosity decreases, it can be seen in Figure 4.11 that the $f_{max}(\alpha)$ shifts to the left. Figure 4.12 shows the mass exponent function, τ(q), against the moment order for

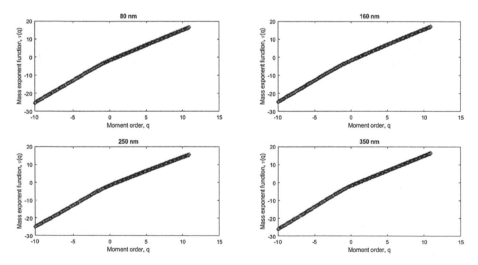

FIGURE 4.12 Mass exponent function (τ(q)) versus moment order (q) for the porous SiO₂ films at different film thicknesses.

the porous structures of SiO_2. It can be seen that all the samples exhibit nonlinear relationships confirming multifractal characteristics of the porosity in the films. From these plots, the values of the slopes were determined as 1.97, 1.97, 1.91, and 2.0 for 80 nm, 160 nm, 250 nm, and 350 nm, respectively. The plots are overlapping and therefore an indication that the function may be insignificant in characterizing the properties of porosity in the films.

4.5 SUMMARY

In the chapter, an overview of defects in structures of thin films has been presented. Most of the discussed defects in the literature of thin films include dislocations, lattice defects, hillocks, and porosity. It is generally agreeable in the literature that defects such as dislocations and lattice defects are difficult to measure since they require powerful measurement tools and equipment. Generally, porosity and hillocks can be imaged through high resolution scanning electron microscopy and atomic force microscopy techniques. Fractal characterizations of hillocks and porosity were presented from published data and shown to be an effective tool in analyzing such defects. Fractal parameters such as fractal dimension, multifractal spectra, generalized multifractal dimension, Hurst exponent, and so forth are important in characterizing the fractal properties of the defects in relationship to the deposition conditions. The fractal theory can be applied on thin film structures to detect the existence of defects or any abnormality on the surface microstructures of thin films.

REFERENCES

[1] B. Şimşek, Ö. B. Ceran, and O. N. Şara, "Difficulties in thin film synthesis," in *Handbook of Nanomaterials and Nanocomposites for Energy and Environmental Applications*, Oxana Vasilievna Kharissova, Leticia Myriam Torres Martínez, and Boris Ildusovich Kharisov, Eds., Cham: Springer International Publishing, 2020, pp. 1–23.

[2] B. Rajeswaran and A. M. Umarji, "Defect engineering of VO2 thin films synthesized by Chemical Vapor Deposition," *Mater. Chem. Phys.*, vol. 245, no. September 2019, p. 122230, Apr. 2020.

[3] X. Xu et al., "Bulk defects induced coercivity modulation of Co thin film based on a Ta/Bi double buffer layer," *J. Magn. Magn. Mater.*, vol. 500, no. January, p. 166388, Apr. 2020.

[4] C. Li et al., "Atomic scale characterization of point and extended defects in niobate thin films," *Ultramicroscopy*, vol. 203, no. February, pp. 82–87, Aug. 2019.

[5] X. You, Y. Huang, Z. Xie, G. Liang, H. Zhu, and Y. Mai, "Ag alloying for modifications of carrier density and defects in Zn-rich (Ag,Cu)2ZnSnSe4 thin film solar cells," *J. Alloys Compd.*, vol. 842, p. 155884, Nov. 2020.

[6] W. Gian, M. Skowronski, and G. S. Rohrer, "Structural defects and their relationship to nucleation of gan thin films," *MRS Proc.*, vol. 423, no. 9, p. 475, Feb. 1996.

[7] W.-R. Liu, W. F. Hsieh, C.-H. Hsu, K. S. Liang, and F. S.-S. Chien, "Influence of the threading dislocations on the electrical properties in epitaxial ZnO thin films," *J. Cryst. Growth*, vol. 297, no. 2, pp. 294–299, Dec. 2006.

[8] M. Nemoz, R. Dagher, S. Matta, A. Michon, P. Vennéguès, and J. Brault, "Dislocation densities reduction in MBE-grown AlN thin films by high-temperature annealing," *J. Cryst. Growth*, vol. 461, no. December 2016, pp. 10–15, Mar. 2017.

[9] M. J. Kappers, R. Datta, R. A. Oliver, F. D. G. Rayment, M. E. Vickers, and C. J. Humphreys, "Threading dislocation reduction in (0001) GaN thin films using SiNx interlayers," *J. Cryst. Growth*, vol. 300, no. 1, pp. 70–74, Mar. 2007.

[10] K. Maehashi, H. Nakashima, F. Bertram, P. Veit, and J. Christen, "Threading dislocation reduction in GaAs films on thin Si substrates," *Phys. E Low-dimensional Syst. Nanostructures*, vol. 2, no. 1–4, pp. 772–776, Jul. 1998.

[11] Y. Tokumoto, N. Shibata, T. Mizoguhci, T. Yamamoto, and Y. Ikuhara, "Atomic structure of threading dislocations in AlN thin films," *Phys. B Condens. Matter*, vol. 404, no. 23–24, pp. 4886–4888, 2009.

[12] X. Zhao, R.-T. Wen, B. Albert, and J. Michel, "Trapping threading dislocations in germanium trenches on silicon wafer," *J. Cryst. Growth*, vol. 543, no. March, p. 125701, Aug. 2020.

[13] R. S. Fertig and S. P. Baker, "Capture cross-section of threading dislocations in thin films," *Mater. Sci. Eng. A*, vol. 551, pp. 67–72, 2012.

[14] J. Bai and S. Wang, "Screw dislocation equations in a thin film and surface effects," *Int. J. Plast.*, vol. 87, pp. 181–203, Dec. 2016.

[15] F. M. Mwema, O. P. Oladijo, S. A. Akinlabi, and E. T. Akinlabi, "Properties of physically deposited thin aluminium film coatings: A review," *J. Alloys Compd.*, vol. 747, pp. 306–323, May 2018.

[16] F. M. Mwema, E. T. Akinlabi, and O. P. Oladijo, "Fractal analysis of hillocks: A case of RF sputtered aluminum thin films," *Appl. Surf. Sci.*, vol. 489, pp. 614–623, Sep. 2019.

[17] H. Barda and E. Rabkin, "Hillocks formation in the Cr-doped Ni thin films: Growth mechanisms and the nano-marker experiment," *J. Mater. Sci.*, vol. 55, no. 6, pp. 2588–2603, Feb. 2020.

[18] T. Arai, A. Makita, Y. Hiromasu, and H. Takatsuji, "Mo-capped Al-Nd alloy for both gate and data bus lines of liquid crystal displays," *Thin Solid Films*, vol. 383, no. 1–2, pp. 287–291, 2001.

[19] S. L. Lee, J. K. Chang, Y. C. Cheng, K. Y. Lee, and W. C. Chen, "Effects of scandium addition on electrical resistivity and formation of thermal hillocks in aluminum thin films," *Thin Solid Films*, vol. 519, no. 11, pp. 3578–3581, 2011.

[20] H.-J. Nam, D.-K. Choi, and W.-J. Lee, "Formation of hillocks in Pt/Ti electrodes and their effects on short phenomena of PZT films deposited by reactive sputtering," *Thin Solid Films*, vol. 371, no. 1–2, pp. 264–271, Aug. 2000.

[21] C. F. Ma, F. Wang, P. Huang, T. J. Lu, and K. W. Xu, "Hillock growth in CuZr metallic glass," *Thin Solid Films*, vol. 589, pp. 681–685, Aug. 2015.

[22] D. Resnik, J. Kovač, M. Godec, D. Vrtačnik, M. Možek, and S. Amon, "The influence of target composition and thermal treatment on sputtered Al thin films on Si and SiO2substrates," *Microelectron. Eng.*, vol. 96, pp. 29–35, 2012.

[23] S. Nazarpour, O. Jambois, C. Zamani, F. Afshar, and A. Cirera, "Stress distribution and hillock formation in Au/Pd thin films as a function of aging treatment in capacitor applications," *Appl. Surf. Sci.*, vol. 255, no. 22, pp. 8995–8999, Aug. 2009.

[24] S. Nazarpour and M. Chaker, "Fractal analysis of Palladium hillocks generated due to oxide formation," *Surf. Coatings Technol.*, vol. 206, no. 11–12, pp. 2991–2997, 2012.

[25] D. Kim, "Effect of SiO2 passivation overlayers on hillock formation in Al thin films," *Thin Solid Films*, vol. 520, no. 21, pp. 6571–6575, Aug. 2012.

[26] S. J. Hwang, J. H. Lee, C. O. Jeong, and Y. C. Joo, "Effect of film thickness and annealing temperature on hillock distributions in pure Al films," *Scr. Mater.*, vol. 56, no. 1, pp. 17–20, 2007.

[27] E. Iwamura, T. Ohnishi, and K. Yoshikawa, "A study of hillock formation on Al Ta alloy films for interconnections of TFT-LCDs," *Thin Solid Films*, vol. 270, no. 1–2, pp. 450–455, Dec. 1995.

[28] B. C. Martin, C. J. Tracy, J. W. Mayer, and L. E. Hendrickson, "A comparative study of Hillock formation in aluminum films," *Thin Solid Films*, vol. 271, no. 1–2, pp. 64–68, Dec. 1995.

[29] A. Karperien, "FracLac for ImageJ, version 2.5," 2014. [Online]. Available: http://rsb.info.nih.gov/ij/plugins/fraclac/FLHelp/Introduction.htm.

[30] J. Schindelin et al., "Fiji: An open-source platform for biological-image analysis," *Nat. Methods*, vol. 9, no. 7, pp. 676–682, Jul. 2012.

[31] R. Shidpour and S. M. S. Movahed, "Identification of defective two dimensional semiconductors by multifractal analysis : The single-layer MoS 2 case study," *Physic. A*, vol. 508, pp. 757–770, 2018.

[32] W. Xu et al., "Hierarchically structured AgO films with nano-porosity for photocatalyst and all solid-state thin film battery," *J. Alloys Compd.*, vol. 802, pp. 210–216, Sep. 2019.

[33] S. Saini et al., "Porosity-tuned thermal conductivity in thermoelectric Al-doped ZnO thin films grown by mist-chemical vapor deposition," *Thin Solid Films*, vol. 685, no. September 2018, pp. 180–185, Sep. 2019.

[34] C. Corbella, "Influence of the porosity of RF sputtered Ta2O5 thin films on their optical properties for electrochromic applications," *Solid State Ionics*, vol. 165, no. 1–4, pp. 15–22, Dec. 2003.

[35] A. Vomiero et al., "Preparation and microstructural characterization of nanosized Mo–TiO2 and Mo–W–O thin films by sputtering: Tailoring of composition and porosity by thermal treatment," *Mater. Sci. Eng. B*, vol. 101, no. 1–3, pp. 216–221, Aug. 2003.

[36] P. Jin, S. Nakao, S. Tanemura, and S. Maruno, "Evaluation of porosity and composition in reactively r.f.-sputtered Ti1 − xZrxN films," *Thin Solid Films*, vol. 271, no. 1–2, pp. 19–25, Dec. 1995.

[37] H. Zegtouf et al., "Influence of oxygen percentage on in vitro bioactivity of zirconia thin films obtained by RF magnetron sputtering," *Appl. Surf. Sci.*, vol. 532, no. July, p. 147403, Dec. 2020.

[38] T. Galy, M. Marszewski, S. King, Y. Yan, S. H. Tolbert, and L. Pilon, "Comparing methods for measuring thickness, refractive index, and porosity of mesoporous thin films," *Microporous Mesoporous Mater.*, vol. 291, no. May 2019, p. 109677, v2020.

[39] C. F. Ramirez-Gutierrez, J. D. Castaño-Yepes, and M. E. Rodriguez-Garcia, "Porosity and roughness determination of porous silicon thin films by genetic algorithms," *Optik (Stuttg).*, vol. 173, no. June, pp. 271–278, Nov. 2018.

[40] R. S. Thomaz, C. R. B. Esteves, and R. M. Papaléo, "Surface morphology and porosity induced by swift heavy ions of low and high stopping power on PMMA thin films," *Nucl. Instruments Methods Phys. Res. Sect. B Beam Interact. with Mater. Atoms*, vol. 435, no. November 2017, pp. 157–161, Nov. 2018.

[41] N.-W. Park et al., "Temperature-dependent thermal conductivity of nanoporous Bi thin films by controlling pore size and porosity," *J. Alloys Compd.*, vol. 639, pp. 289–295, Aug. 2015.

[42] İ. A. Kariper, "Production and characterization of Tel x (x: 2, 4) thin films: Optical, structural properties and effect of porosity," *Mater. Des.*, vol. 106, pp. 170–176, Sep. 2016.

[43] D. Fu and X. Cheng, "Exploring the effect on the columnar structure and porosity of the synthesized Be films by oblique angle deposition in magnetron sputtering," *Phys. B Condens. Matter.*, vol. 590, no. April, p. 412221, Aug. 2020.

[44] A. A. Sycheva, E. N. Voronina, T. V. Rakhimova, and A. T. Rakhimov, "Influence of porosity and pore size on sputtering of nanoporous structures by low-energy Ar ions: Molecular dynamics study," *Appl. Surf. Sci.*, vol. 475, no. September 2018, pp. 1021–1032, May 2019.

[45] F. Oudrhiri-Hassani, L. Presmanes, A. Barnabé, and P. Tailhades, "Microstructure, porosity and roughness of RF sputtered oxide thin films: Characterization and modelization," *Appl. Surf. Sci.*, vol. 254, no. 18, pp. 5796–5802, Jul. 2008.

[46] K. P. Mogilnikov, Dongchen Che, M. R. Baklanov, Kangning Xu, and Kaidong Xu, "*Review of thin film porosity characterization approaches,*" in *2017 China Semiconductor Technology International Conference (CSTIC)*, Shanghai, China. 2017, pp. 1–4.

[47] T. Galy, M. Marszewski, S. King, Y. Yan, S. H. Tolbert, and L. Pilon, "Comparing methods for measuring thickness, refractive index, and porosity of mesoporous thin films," *Microporous Mesoporous Mater.*, vol. 291, no. August 2019, p. 109677, Jan. 2020.

[48] N. Kumar, P. Sanguino, S. Diliberto, P. Faia, and B. Trindade, "Tailoring thin mesoporous silicon-tin films by radio-frequency magnetron sputtering," *Thin Solid Films*, vol. 704, no. March, p. 137989, Jun. 2020.

[49] F. Chen and A. H. Kitai, "Growth of nanoporous silicon dioxide thin films using porous alumina substrates," *Thin Solid Films*, vol. 517, no. 2, pp. 622–625, Nov. 2008.

5   Mono-Fractal Analyses of Roughness of Sputtered Films

5.1 INTRODUCTION

Sputtering is a random process (as detailed in Chapter 1). As such, achieving a uniform coverage of the substrate surface and high quality of the sputtered thin films would require taking into consideration two important points:

  i. Accurate fine-tuning of the sputtering process via adjusting the parameters and
  ii. Detailed analyses of surface roughness of the sputtered thin films

The first point on the fine-tuning of the process parameters was described in Chapter 1 and it requires experience and iterative studies to determine the optimum conditions for the specific sputtering problem. The second aspect of studying the surface roughness helps in assessing the quality of the grown thin films in relation to the process parameters for continuous quality improvement of the sputtering technology. Additionally, roughness characterization can assist on developing bio-inspired or patterned thin films for specific applications. Of the roughness methods, the use of fractal theory has been shown to provide a detailed understanding on the mechanism of thin film growth during the sputtering or other deposition processes. As it was stated in Chapter 2, it is possible to understand the lateral evolution of the structures during thin film sputtering. As such, there are so many studies utilizing fractal theory in understanding the roughness of thin films [1–6].

In this chapter, the application of fractal theory on studying the surface roughness of sputtered thin films is presented with emphasis on mono-fractal algorithms described in Chapter 3. The chapter uses the results from the authors of this book and published literature to illustrate these concepts.

5.2 VERTICAL AND LATERAL ROUGHENING IN SPUTTERING

The surface roughness greatly influences the properties and functionality of thin film components used in various fields. Some of the properties of thin films dependent on surface topography and roughness include electrical and electronics responses, optical, thermal, adhesion, friction, hydrophobic, and mechanical stability [7]. The study of surface properties of thin films is very important for the purposes of quality control and enhancement of their functionality. There are several standards of undertaking surface roughness measurement and some of them include ISO 25178, and ASTM D7127-13 [8]. The most common equipment for measurement of surface topography includes:

  i. X-ray reflectometry (XRR)
 ii. Scanning tunneling microscopy (STM)
iii. Atomic force microscopy (AFM)
 iv. Optical non-contact profiler
  v. Laser interferometry
 vi. Scanning electron microscopy (SEM)

In thin films, the AFM technique is preferred over the other techniques for surface topography due to the following reasons [9]:

  i. It has low vertical noise and high lateral resolution.
 ii. It can capture images at the atomic scale since it is possible to resolve features of less than 1 nm.
iii. The AFM facility is versatile and flexible and besides the topographic results, it can give other properties on adhesion, temperature, elasticity, and conductivity of the thin films.
 iv. The versatility of the machine also enables them to image a wide range of materials, including metals, and polymer, and the state of materials such as powders, liquid, solid, hydrogel, and so forth.
  v. The images can be undertaken in different conditions such as air, vacuum, or liquid.

The AFM provides images on the surface topography of the thin films and these images are then analyzed for surface roughness. In most cases, the average roughness (Ra) and root mean square roughness (Rq) are used to describe the surface roughness from the AFM results (Ra and Rq are statistical parameters). The equations for these parameters are readily available in the literature and are usually implemented in software attached to the AFM facilities or other software such as Gwydion. The parameters Ra and Rq represent the vertical roughness of the surfaces of the thin films and have been extensively used in literature to study surface topography [10,11]. However, as discussed by Mwema et al. [8], the use of these parameters for surface topography characterization is limited by the following factors:

  i. These parameters do not take into consideration the lateral deviation characteristics of the surfaces of the thin films.
 ii. The Ra and Rq parameters do provide a distinction between valleys and peaks, thus, two surfaces can have the same value of Rq and Ra but have very different surface topographies.
iii. These parameters are greatly influenced by the AFM measurement conditions such as the sampling interval, scanning scales, and other specific aspects of the AFM measurement. Such factors increase the chances of error in the roughness values.

To overcome the above challenges associated with the statistical approaches (Ra and Rq), fractal techniques are used. These techniques provide information on the lateral roughness, differentiates between peaks and valleys, and the results are not influenced by the measurement conditions of the AFM facility. Through fractal methods, roughness is characterized by fractal dimension, Hurst exponent, equivalent roughness, spectral length, and among others. As such, fractal methods are effective tools for the characterization of lateral evolution of surface features on thin film surfaces.

In sputtering, a complete surface roughness characterization, both vertical/ statistical and lateral, is important. This is because there is a growth of film's structures in both directions due to the random nature of the process. The continuous dislodgement of target material and deposition onto the substrate surface leads to vertical development of the films whereas due to the fact that there is a continuous rotation of the substrate, there occurs lateral evolution of the structure of the films. A typical model leading to vertical and lateral roughening has been presented by the authors of this book for Al thin films grown via magnetron sputtering on steel substrates [12]. It can be deduced that vertical roughening is responsible for thickness evolution whereas lateral roughness is directly related to the surface coverage of the substrate.

5.3 THE RELATIONSHIP BETWEEN LATERAL AND VERTICAL ROUGHNESS IN SPUTTERING OF METALLIC/ALLOY THIN FILMS

In this section, several case studies showing the relationship between the vertical and lateral roughness of sputtered thin films are illustrated. The case studies are based on original results by the authors and those published in literature by other authors. The case study 1 is presented as a form of a research article including detailed experimental and measurement strategy as a step-by-step guide for the readers. The descriptions of the results based on the methodology are presented and the relationships between the fractal and vertical roughness characteristics are detailed. The case studies 2, 3, and 4 are based on published literature and the reported results are presented here to show the relationship between the fractal and vertical roughness parameters in thin film sputtering processes. Finally, an informed conclusion on the relationship between lateral roughness based on the fractal dimension and vertical roughness based on statistical parameters is presented.

5.3.1 CASE 1: TOPOLOGY AND FRACTAL ANALYSIS OF THIN ALUMINIUM FILMS GROWN BY RADIO-FREQUENCY MAGNETRON SPUTTERING

Introduction

In this case, the vertical and lateral surface properties of Al thin films deposited on mild steel substrates through radio-frequency (RF) magnetron sputtering at varying low substrate temperatures (44°C, 80°C, and 100°C) were studied using atomic force microscopy (AFM). Topology and fractal studies were undertaken on the surface micrographs to explore the structural properties of the films. The interface width, average roughness, order-based statistical features, autocorrelation, height–height correlation, power spectral density (PSD), Minkowski functionals (MFs), lateral correlation length, fractal dimension, and roughness exponent were computed. Furthermore, a numerical study using computer-generated mounded surfaces was undertaken to explain the AFM results. The analysis shows that varying the substrate temperature, even at very low ranges, affects the evolution of the surface structures. The fractal dimension (D) is shown to increase with the increase of the substrate temperature and the numerical study has revealed that the Al thin films deposited through RF magnetron sputtering exhibit both self-affine and mounded surface properties. The microstructures obtained through field emission scanning electron microscopy (FESEM) and X-ray diffraction (XRD) confirmed the fractal analyses presented here.

Background

Aluminium thin films prepared by sputtering are used as an interconnection material in integrated circuits (ICs) due to their excellent adhesion to silicon and high electrical conductivity. However, the deposition of Al films on Si is associated with some difficulties because of the low melting temperature of Al (660°C) and low eutectic point with Si (570°C) [13]. Additionally, Al deposited on non-metals (such as Si and glass) experience the formation of the high density of hillocks and have poor resistance to electromigration. To eliminate some of these limitations, a barrier metal is introduced between the substrate (e.g. Si) and the Al films [14]. In such cases, the barrier metal serves as the substrate for the films and the microstructure of the film growth significantly changes [13,15].

The surface characteristics such as roughness of thin films have been reported to influence their electrical, optical, hydrophobicity, and other physicochemical properties [16,17]. As such the understanding of morphological evolution at the surfaces of thin films is very important for the enhancement of the deposition processes. Statistical and fractal analyses have been used extensively to study the surface properties of various thin films [18,19]. Yadav et al. have extensively used fractal analysis to compute roughness exponent, lateral length, and fractal dimensions of LiF and ZnO films [20–22] and reported that morphological properties strongly depend on the parameters of the deposition technique and film thickness. Buzio et al. [18] employed the height–height correlation function (HHCF) to determine the roughness characteristics of self-affine thin films of carbon. The dynamic scaling analysis of silver thin films has also been undertaken through fractal analysis tools (PSD and HHCFs) [19]. In a recent study, Liu et al. [23] have used the fractal concepts to analyze the topographical properties of graphene nanosheets. The micromorphology of ITO prepared by sputtering has also been analyzed with areal autocorrelation functions [24]. From the fractal analysis, researchers have been able to extensively explain the roughness properties of various film surfaces and some of these characteristics include interface width, lateral correlation length, roughness exponent, fractal dimension, and spatial patterning of the structural units. From these analyses, contrasting relationships among these characteristics have been reported. For instance, in the study [22], the fractal dimension was shown to increase with interface width whereas vice versa was reported in the study [21]. This suggests that for different types of films, the fractal parameters exhibit different relationships. Despite being such an important thin film for IC applications, the fractal analysis of the sputtered Al has not been sufficiently reported. More specifically, the fractal analysis of thin aluminium films with varying low substrate temperature during sputtering is lacking in the literature. It is expected that the fractal (roughness scaling) behavior of Al films changes with the substrate temperature. In this work, therefore, topological and fractal analyses of sputtered Al thin films are presented.

Experimental Details

Aluminium thin films were prepared at varying low substrate temperatures of 44°C, 80°C, and 100°C on mild steel substrate through the RF magnetron sputtering process in high vacuum conditions. The sputtering was undertaken for 2 hours at a constant RF power of 200 W. The substrate surfaces were ground to 1 μm surface finish, washed in distilled water, and rinsed in acetone. A high-purity, 99.999% aluminium from HHV Ltd, United Kingdom, was used as the target material. After sputtering, the samples were cooled slowly inside the vacuum chamber of the sputtering system and then sliced into 10×10 mm^2 for AFM measurements. The AFM imaging was undertaken in Veeco Dimension 3100 machine in tapping mode at a scan area of 1×1 μm^2. A detailed description of the sputtering and AFM measurements was presented earlier [10].

For statistical accuracy, 10 AFM images were taken across the surface of each sample and their average (arithmetic means) values were reported. Statistical analysis between the roughness values from different images for the same surfaces was performed using IBM SPSS for windows (trial version). One-way ANOVA was used to determine the differences and P-values were shown to be more than 0.05 indicating statistically insignificant differences among each of the 10 measurements per surface. The obtained AFM digital data (256×256 pixels) were analyzed for topology and fractals using Bruker Nanoscope (version 530r3sr3) and Gwyddion™ [25] software. FESEM and XRD were undertaken to validate the fractal description of the AFM data.

Analytical Methods

Roughness Characteristics

The roughness characteristics of surfaces, also known as moment-based quantities, from AFM imaging are used to quantify the height distribution of surface features and usually computed using height function integral, $z(i,j)$. Some of these parameters are the root mean square, average roughness, skewness, and coefficient of kurtosis and were used earlier to describe the topography of thin aluminium films [10]. The average roughness, R_a, and interface width, commonly referred to as root mean square roughness (w), are two main quantitative parameters used to describe the surface morphology of thin films. Mathematically, these parameters are defined as follows (Equations 5.1 and 5.2).

$$R_a = \frac{1}{m^2} \sum_{i,j=1}^{m} \left| z(i,j) - \langle z(i,j) \rangle \right| \qquad (5.1)$$

$$w = \frac{1}{m}\sqrt{\sum_{i,j=1}^{m}\left(z(i,j)-\langle z(i,j)\rangle\right)^{2}} \tag{5.2}$$

The function $z(i,j)$ represents the height of the surface features from the AFM micrograph at coordinates (i,j) whereas $\langle z(i,j)\rangle$ denotes the average height of the surface over $m \times m$ points.

Skewness, R_{sk}, describes the symmetry of the distribution of surface features and it is given by Equation (5.3).

$$R_{sk} = \frac{1}{m^{2}w^{3}}\sum_{i,j=1}^{m}\left(z(i,j)-\langle z(i,j)\rangle\right)^{3} \tag{5.3}$$

From Equation (5.3), m represents the root mean square roughness. For peak and valleys dominant surfaces, the values of R_{sk}, is positive and negative, respectively. Additionally, skewness is zero for symmetric distributions, positive for right-tail dominant distributions and negative for left-tail dominant distributions [26].

The measure of spikiness in the distribution of surface profiles in AFM is described by the coefficient of kurtosis determined shown in Equation (5.4).

$$R_{ka} = \frac{1}{m^{2}w^{4}}\sum_{i,j=1}^{m}\left(z(i,j)-\langle z(i,j)\rangle\right)^{4} \tag{5.4}$$

If $R_{ka}=3$, the surface is said to have a Gaussian distribution, $R_{ka}>3$ the surface is spiky and bumpy if $R_{ka}<3$ (Aqil et al., 2017; Mwema et al., 2018).

Autocorrelation Function

Autocorrelation function (ACF) describes the dependence of the signal on its own at different time shifts. In the surface analysis, ACF has two uses as earlier stated by Yadav et al. [21]: (1) To identify the suitable spatio-temporal model in non-random AFM data and (2) detect non-randomness in the surface data of surfaces. For discrete AFM data, the autocorrelation function $(A(r))$ along the direction of fast scan (x-direction) is expressed in terms of height function $z(i,j)$ as shown in Equation (5.5) [21,27].

$$A(r = ld) = \frac{1}{m(m-l)w^{2}}\sum_{j=1}^{m}\sum_{i=1}^{m-l}z(i+l,j)z(i,j) \tag{5.5}$$

Where d denotes the lateral distance between any two adjacent pixels, l is the immediate pixel before the point (m) under consideration.

Height–Height Correlation Function

HHCF determines the power differences between points and is used to differentiate between self-affine and mounded characteristics of surfaces of thin films [22]. The discrete one-dimensional HHCF $H(r)$ of $m{\times}m$ area of AFM micrograph along the fast scan direction (x-direction) is computed as shown in Equation (5.6) [21].

$$H\left(r=ld\right)=\frac{1}{m\left(m-l\right)}\sum_{j=1}^{m}\sum_{i=1}^{m-l}\left[z\left(i+l,j\right)-z\left(i,j\right)\right]^{2} \tag{5.6}$$

HHCF and ACF are related by the following relationship (Equation 5.7).

$$H\left(r\right)=2w^{2}\left[1-A\left(r\right)\right] \tag{5.7}$$

The $H(r)$ profile reveals two important regimes depending on the magnitude of r and correlation length τ. Lateral correlation length (τ) is defined as the magnitude of r when the HHCF is equal to $(1-1/e)$. This occurs at very large values of r. For self-affine surfaces, $\tau \ll r$, $H(r) \propto 2w^2$ and $\tau \gg r$, $H(r) \propto H_0 r^{2\alpha}$. For very small values of r, the roughness exponent, α, is directly related to the fractal dimension (D) of the surfaces as $D=3-\alpha$ and this applies only if α less than unity [21,28].

The physical interpretation of the $H(r)$ profile employs curve fitting tools for complex surfaces such as those of thin aluminium films. In this work, we used a power law fitted at small values of r and Gaussian functions fitted at large values of r to compute roughness exponent, lateral correlation length, and fractal dimension. The Gaussian function best-fit curve used is written as shown in Equation (5.8) as reported in the literature [21,22].

$$H\left(r\right)=2w^{2}\left[1-e^{-\left(\frac{r}{\tau}\right)^{2a}}\right] \tag{5.8}$$

Power Spectral Density

PSD indicates the distribution of powers of various signals over a certain spectrum of frequency. Besides describing the fractal behavior of thin film surfaces, PSD function (PSDF) separates the contributions of various length scales to the surface roughness. In the log-log plot of PSDF, the start of the self-affine behavior is indicated by a sharp knee. PSDF for thin Al film surface analysis was detailed in our earlier article [29]. The one-dimensional discrete PSDF of rows is written as in Equation (5.9) [27].

$$PSDF=\frac{2\pi}{mlz}\sum_{j=0}^{l-1}\left|P_{j,l}\right|^{2} \tag{5.9}$$

Where P is the discrete Fourier coefficient of the rows (e.g. jth row)

Minkowski Functionals

MFs are used to describe the morphology and shape of stochastic models in two-dimensional Euclidean space [25]. They are also used to mathematically discriminate structures of different geometries which may otherwise reveal the same fractal dimension. This functionals also analyze spatial features of random fields, to differentiate between different models, as well as to evaluate experimental data for various models. In two-dimensional space there exist three MFs namely volume, V, surface, S, and Euler-Poincaré characteristic (also known as connectivity), ξ. These MFs are computed as follows (Equation 5.10) according to the number of pixels above the threshold (white), N_{white}, black (pixels above the threshold), and intermediate pixels (black-white boundary pixels), N_{bound}.

$$V(z) = \frac{N_{white}}{N}, S(z) = \frac{N_{bound}}{N}, \xi(z) = \frac{C_{white} - C_{black}}{N} \qquad (5.10)$$

Where N represents the number of pixels on the analyzed area and C is the number of continuous sets of pixels.

Results and Discussion

Vertical Roughness Characteristics

The visual surface morphologies of the aluminium thin films deposited on mild steel at various substrate temperatures (T_s) as characterized by AFM are shown in Figure 5.1. The histograms of the counts of the surface features (structures) from each of the AFM micrographs are shown in Figure 5.1(d). As shown, the size and cumulative height distribution of surface features vary with the deposition temperature. The structures at 44°C are the smallest with few counts of surface events compared to the other temperatures. The interface width, w, and average roughness, R_a,

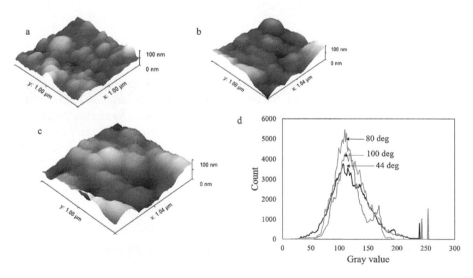

FIGURE 5.1 The representative AFM micrographs of aluminium thin films grown at a substrate temperature of (a) 44°C (b) 80°C, and (c) 100°C, respectively (d) A histogram of the count of the number of data points (structures) of the surfaces at various Gray values (local height) (a)–(c).

TABLE 5.1

Average Surface Statistical Quantities (z(i,j)-Based Features) of Aluminium thin Films from AFM Microscopy at Various Deposition Temperatures

Surface Quantity	44°C	80°C	100°C
RMS roughness (W), nm	14.17	16.40	18.92
Mean roughness (R_a), nm	11.14	12.64	15.44
Skew (R_{sk})	0.48	0.83	0.04
Kurtosis (R_{ka})	0.48	0.51	−0.56
Max. peak height (z_p), nm	52.76	58.75	50.87
Max. pit depth (z_p), nm	47.24	46.25	38.13

skewness, R_{sk}, and coefficient of kurtosis, R_{ka}, computed according to formulae (Equation 5.1 to Equation 5.4) from the AFM images are summarized in Table 5.1. Other significant surface parameters such as median height, maximum pit and depth, peak height variation, and entropy are also reported. We observe that w and R_a increase (generally) with an increase in deposition temperature. All the surfaces are bumpy and consist of valleys and peaks according to the skewness and kurtosis results – the kurtosis values are less than 3 and skewness values are all positive [26]. These observations can be attributed to the agglomeration of the adatoms into larger particles (or clusters) on the surface of the substrate due to extra energy provided by the higher substrate temperature [22,30] during the continuous film stage of the sputtering process. The maximum peaks and pits in Table 5.1 indicate the presence of vertically large structures and pores/defects on the aluminium films. At 80°C, the highest values of peaks and pits are observed, which could be the reason for the high counts of events observed in the histogram in Figure 5.1(d). It means that the films are growing vertically at this temperature without adequately flowing through the pores. The lowest values of maximum pits and peaks are observed at 100°C, which means that the target atoms have started flowing laterally on the substrate surface to fill the pores of the films.

Fractal Characteristics

The autocorrelation $A(r)$ and height–height correlation $H(r)$ functions of the Al films at each substrate temperature were computed using methods in Equations 5.5 and 5.6 and the results plotted in Figure 5.2 and Figure 5.3 respectively. Both $A(r)$ and $H(r)$ functions are used to examine the self-affine and fractal characteristics of thin film surfaces. $A(r)$ functions are traditionally used to show the repetitive characteristics of surfaces [24,31]. For all the surfaces, the $A(r)$ decays monotonically (up to $A(r)=−1.0$) with increasing shift (r) as shown in Figure 5.2(b). As shown, the surfaces of the thin films grown at 44°C have the shortest length of decay whereas surfaces at 100°C have the longest decay length. Whereas the decay limit may be approximately $A(r)=−1.0$ for 44°C and 100°C, it is larger for 80°C. In $A(r)$ analysis, the decay length illustrates the anisotropic ratio, which describes the uniformity of surface textures. In this case, the shortest decay length is an indication of significantly uniform

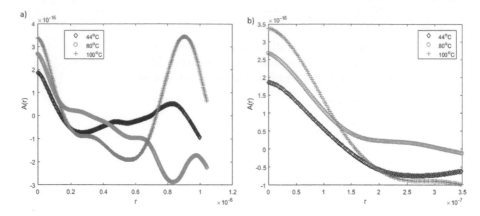

FIGURE 5.2 (a) Autocorrelation function **A(r)** of thin aluminium films grown at 44°C (◊), 80°C (o), and 100°C (+). (b) Zoomed correlation regions of the **A(r)** functions at small values of **r** (r<3.5×10⁻⁷) plotted in Figure 5.2(a). There is nearly an exponential decay in the **A(r)** as the values of **r=ld** increases.

surface texture at low deposition temperatures [22]. The value of shift (r) when A(r) is equal to the inverse of the natural logarithm (~**0.3679**), is the lateral correlation length (τ). From Figure 5.2, we determine that τ is 108.20 nm, 121.06 nm, and 130.61 nm for the Al films grown at 44°C, 80°C, and 100°C, respectively. We observe in Figure 5.2(a) that beyond shift, r≈**0.5**, the A(r) tends to exhibit random oscillatory behavior for all the samples (the oscillations have different amplitudes and frequencies), depicting mounded surfaces in the Al thin films.

Height–height correlation function ($H(r)$), on the other hand, shows the interrelationships among the surface features in the lateral distribution. It is related to $A(r)$ as $H(r)=2w^2 (1-A(r))$. The double log plots of the computed $H(r)$ for the Al thin films shown in Figure 5.3 depict self-affine characteristics and follow an allometric scaling law, $H(r)=H_0 r^{2\alpha}$, where, H_0 is known as the pseudo-topothesy. For all the samples, the plots consist of two distinct regions: (i) A linear region at small values of r, and (ii) a plateau region at the large magnitude of the shift (r). To derive the fractal characteristics from $H(r)$, we fitted the linear region to a power law and the latter region with Gaussian function (as described earlier). The results of the curve fitting are summarized in Table 5.2. We observe that the AFM data fitted accurately to a power law (not shown in Figure 5.3) and Gaussian function at small and large values of the shift (r) according to the R^2 values in Table 5.2 and Figure 5.3 for all the Al thin film surfaces. The transition point between the two regions shows the lateral correlation length (τ) of the $H(r)$ and this parameter determines the horizontal distance at which the height quantities of the surface features are no longer related – it indicates the transition from pits to valleys/peaks. The average values of correlation length (τ) from the correlation function for all the samples are shown in Table 5.2 and we can see that the lowest value is obtained at the lowest sputtering temperature (44°C). It, therefore, means that we move a shorter distance from the correlated surface features to independent structures (at low substrate temperature), indicating higher lateral

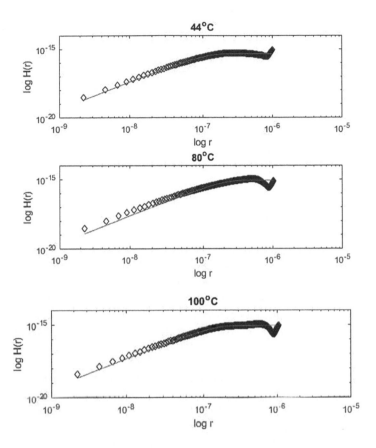

FIGURE 5.3 Double log plots of $H(r)$ for the thin aluminium films deposited at different substrate temperatures. The continuous line in each of the plots indicates the Gaussian best fit according to Equation 5.8 for larger values of r (nm). The power law was fitted on the correlated regions, at values of r, (not shown).

TABLE 5.2
Average Curve-Fitting Parameters of the HHCFs According to Section 3.3 and Computation of Fractal Dimension from HHCF (D_{HHCF})

Samples	R^2 (Correlation Region)	τ (nm)	α	$D_{HHCF}=3-\alpha$
44°C	0.9861	105.29±0.002	0.7917±0.00024	2.2083
80°C	0.99651	123.16±0.003	0.8640±0.00175	2.1360
100°C	0.99188	134.55±0.001	0.8817±0.00045	2.1083

roughness at lower substrate temperature. Higher values of lateral lengths are observed at 80°C and 100°C, indicating the growth of similar surface features. The roughness exponents (α) are computed from the curve-fitting in the linear region and are also shown in Table 5.2. The magnitude of roughness exponent reveals the extent of development of the surface structures; larger values indicate smooth and developed surfaces whereas smaller values correspond to rugged (rough) film surfaces where rugged surfaces correspond to porosity-prone films [21,22].

The results show that the highest exponent occurs at 100°C and the lowest at 44°C. We further observe that there is no distinct relationship between the correlation length, τ, and roughness exponent α of Al thin films in the current study. Similar observations were presented in an earlier height–height autocorrelation analysis of LiF films by Yadav et al. [21]. From the height–height correlation results, the fractal dimension D was determined according to Equations 5.6 to 5.8 and the values tabulated in Table 5.2. Fractal dimension indicates the complexity of surface morphologies in thin films [28].

Because of its importance in describing self-affine surfaces, fractal dimension (D) was further determined using other methods besides the $H(r)$ function. The other techniques used are box counting (cube method), triangulation, and PSD methods [32]. Figure 5.4 shows the double log plots of PSDF versus the spatial frequency of all the samples. As observed in Figure 5.4, all the PSDF curves have three distinct regions namely plateau or white noise region at a low spatial frequency, knee, and constant power regions. This is a characteristic of self-affine surfaces. As extensively described in our earlier study [10], the fractal dimensions have been determined from the PSDF plots by fitting a power law (not indicated in Figure 5.4), the reader is referred to articles [33, 34] within the knee region and the results are presented in Table 5.3 for all the deposition temperatures. Figure 5.5 illustrates the computation of the fractal dimension (D) using the cube counting technique. The method uses double plots of the number of pixels (N) versus length scale (or lattice constant) and then computing the slope of the best-fit curve as the fractal dimension. In all the cases, we have found that the straight-line curve provides the most accurate fit (with all correlation value $R^2 > 0.9$) of the cube counting data. Using a similar procedure, we computed the fractal dimension using the triangulation method and all the results are tabulated in Table 5.3.

From Table 5.3, we observe that from the four fractal dimension determination methods (HHCF, cube counting, triangulation, and PSDF) values of D obtained ranged between 2.0 and 2.5. As reported in the literature, most thin films deposited through sputtering exhibit significant self-affine characteristics, and since D of such surfaces vary between 2 and 3, it means that the four methods provide a good approximation of the fractal dimension of Al thin films. A similar approach in which D was computed through different methods has been used in literature for graphene nanostructures (Levchenko et al., 2016). To establish the relationship between the deposition temperature and fractal dimension of our thin films, we averaged the fractal dimensions obtained by the four techniques for each temperature and the result plotted in a similar way to an earlier study [35]. A linear function fit ($T_s(D) = 1770.7D\text{-}3824.2$) of the data shows that the fractal dimension of the sputtered Al films increases with substrate temperature although this assertion is at a

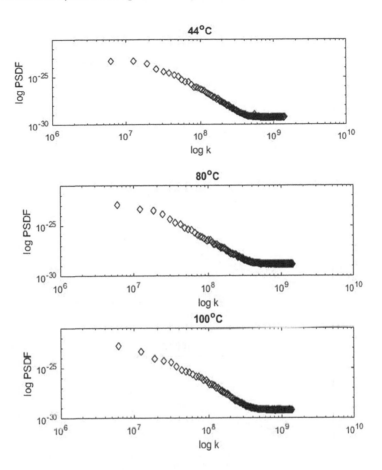

FIGURE 5.4 log-log plots of **PSDF** versus spatial frequencies (k, m^{-1}) of thin Al films deposited at different substrate temperatures. PSDF is in one dimension (m^3).

TABLE 5.3

Fractal Dimensions of the Sputtered thin Al Films Determined by different Computation Methods

Method	44°C	80°C	100°C
HHCF	2.2083±0.0003	2.1360±0.00025	2.1083±0.00012
Cube counting	2.1490±0.00029	2.1190±0.00038	2.1102±0.00025
Triangulation	2.2090±0.00018	2.1470±0.00013	2.1350±0.000101
PSDF	2.1790±0.00031	2.1390±0.00027	2.1022±0.00016
Average D	**2.1863±0.0003**	**2.1353±0.0015**	**2.1139±0.0001**

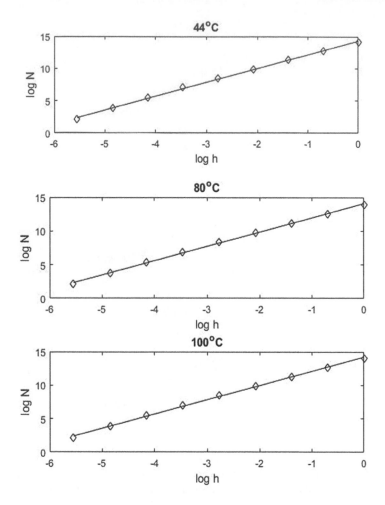

FIGURE 5.5 log plots of N(h) versus length scale (box size) (h) based on cube counting method for determination of fractal dimension, D. The solid line representing the curve fitting on the actual data is represented by ◊ markers. The R^2 values corresponding to the 44°C, 80°C, and 100°C curve fitting are 0.99836, 0.99887, and 0.99837, respectively.

much less accuracy based on the $R^2 = 0.89144$ value for the curve fitting. Similar observations were presented in the literature by Yadav et al. [21] where D for LiF deposited through electron beam evaporation was shown to increase with the thickness of the film. However, contrary results were reported for LiF deposited through electron beam evaporation on a silicon substrate at varying substrate temperature [22]. For ZnO films prepared through electron beam evaporation, the fractal dimension was shown to have a nonlinear relationship with the angle of deposition [20]. In this study [20], however, it was noted that the largest value of D corresponds to jagged (rough) surfaces while smaller values indicate locally smooth surfaces. Similar results were reported on silver films deposited through the vacuum vapor deposition method [35]. However, contrary results were presented on fractal analysis of organic coatings by Talu et al. [36] in which large values of D corresponded to small roughness values. Although our results suggest a significant relationship between D and T_s, we suggest further studies to confirm relationships among roughness, fractal dimension, and deposition temperature.

Minkowski Functionals

Figures 5.6 to 5.8 show graphs of two-dimensional MFs, $V(z)$, $S(z)$, and $\xi(z)$, respectively. These functionals are used to describe the global properties of 3D patterns corresponding to the quantity of structural units to explain the relationship between the surface properties and topology [37,38]. The Minkowski volume ($V(z)$) in Figure 5.6 illustrates the number of pixels with a height threshold larger than height function ($z(i,j)$) of the surface features [31] and as shown, at all the deposition temperatures, the volume decreased slowly from 1.0 to 0.95 between $z = 0$ and 25 nm and rapidly from 0.95 to 0.2 between $z = 25$ and 75 nm. Beyond $z = 75$ nm, the volume decreased slowly again to 0.

The $V(z)$ of Al thin films deposited at 80°C and 100°C reveal clearer symmetries about $V(z) = 0.5$, $z = 45$ nm, and $V(z) = 0.5$, $z = 40$ nm, respectively, as compared to those deposited at 44°C whose symmetry ($V = 0.5$) occur at $z = 50$ nm. The lowest symmetry observed at low substrate temperature is an indication of uneven distribution of the structure of the film. At 100°C, there is the formation of large, even, and continuous films on the surface of the substrate leading to asymmetrical Minkowski volume in Figure 5.6.

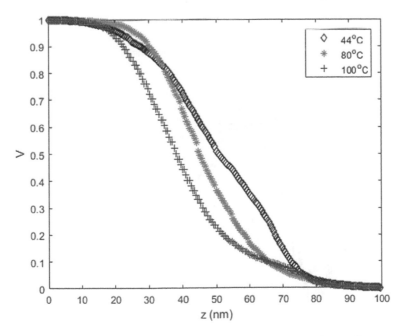

FIGURE 5.6 The Minkowski volume, V (unitless), for the thin aluminium films at various deposition temperatures

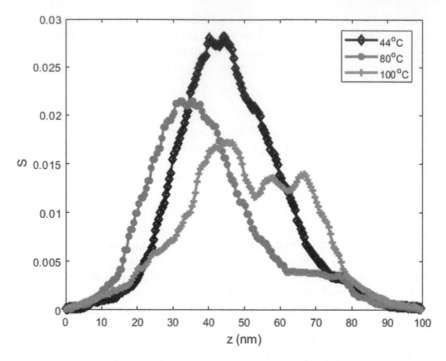

FIGURE 5.7 The Minkowski surface, S (unitless), for thin aluminium films at different deposition temperatures.

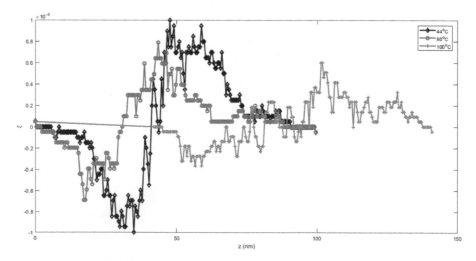

FIGURE 5.8 The Minkowski connectivity (ξ, unitless) for the thin aluminium films at different deposition temperatures.

The Minkowski boundary length (S(z)) graphs shown in Figure 5.7 for all the samples reveal that S(z) increases gradually from zero to maximum and then falls to zero. Al films prepared at 44°C and 80°C exhibit Minkowski surface densities with narrow peaks and Gaussian distribution whereas samples prepared at 100°C reveal asymmetrical and distributions consisting of broad peaks with oscillations at maximum S. We observe that the maximum values of Minkowski boundary lengths are **0.028, 0.023,** and **0.017** for Al films deposited at 44°C, 80°C, and 100°C, respectively. Lower values of maximum S depict laterally developed, flat and continuous surfaces, indicating the growth of structures at 100°C than at 44°C [39]. Furthermore, a broad peak of the S observed at 100°C is an indication of the dominance of valleys over pits (pores and defects) at this substrate temperature.

The Minkowski connectivity graphs shown in Figure 5.8 $\xi(z)$ are theoretically used to describe the connectivity of spatial patterns which are sensitive to shifts in morphologies and geometries of thin film surfaces [40]. The Minkowski connectivity profiles at 44°C and 80°C exhibit oscillatory behaviors (nearly constant amplitude) with minimum values (maximum density of valleys) of $\xi(z)$ as **−0.98×10⁻⁴** (at z=35 nm) and **−0.75×10⁻⁴** (z=25 nm), respectively, and maximum values (maximum density of peaks) as **0.95×10⁻⁴** (z=48 nm) and **0.75×10⁻⁴** (z=45 nm), respectively. We also observe that the curve for 100°C exhibits oscillatory form with exponentially growing amplitude. The maximum values for 100°C curves are obtained as **0.6×10⁻⁴** (z=100 nm) whereas the minimum values are determined as **−0.28×10⁻⁴** (z=50 nm) as seen in Figure 5.8. The largest part of the curve at 100°C is dominated by positive values indicating the dominance of valleys over pits on the surface.

Based on the presence of both positive and negative values of Minkowski connectivity, it can be deduced that at all the deposition temperatures, there is the presence of valleys and pits in the Al thin films prepared through magnetron sputtering. Furthermore, large parts of all the Minkowski connectivity curves (shown in Figure 5.8) are within the positive region indicating that these surfaces are generally rough. It has been reported elsewhere that Al thin films deposited on metallic substrates generally have high roughness as compared to those deposited on glass, polymers, and semiconductors [41]. Additionally, the oscillatory behavior of the Minkowski connectivity curves presented here is an indication that the surface valleys and peaks are randomly distributed and not interconnected to each other. It, therefore, means that during sputtering, time and the rotational speed of the sample holder should be optimally selected to enhance the random deposition of the target atoms onto the substrate for effective 'close-up' of pits and reduction of the surface roughness. The Minkowski connectivity profile at 100°C is dominated by positive values confirming this observation.

To understand the physical reason for the evolution of roughness of topological and fractal properties of Al thin films during sputtering with varying deposition temperatures, we have extracted diagonal line profiles from the AFM micrographs in Figure 5.1. The line profiles are shown in Figure 5.9 (the inset on the lower left of the figure shows the direction and location of the extracted profiles). The profiles at 44°C seem highly serrated and the serrations reduce as the substrate temperature increases. However, the heights of the surface features are not directly correlated to

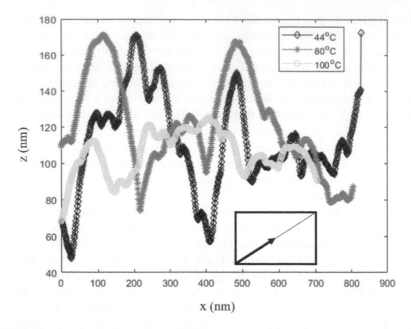

FIGURE 5.9 Section profiles of the AFM micrographs of the Al films sputtered at varying low substrate temperatures.

the substrate temperature. It, therefore, implies that the increase in substrate roughness of Al films deposited by RF magnetron sputtering can be attributed to the growth of lateral surface features rather than vertical (perpendicular to the substrate) growth. These results can further be related to the fractal dimensions obtained in Table 5.3.

Numerical Study

To decipher the fractal behavior of our Al thin films deposited through RF magnetron sputtering, we have analyzed a computer-generated mounded surface shown in Figure 5.10. Our approach is based on a previous study by Siniscalco et al. [42] in which it was stated that most thin films exhibit mounded surfaces. The model is shown in Figure 5.10 and it generated based on the observations made earlier in the AFM images. The model surface is dominated by bumps and valleys with few pits and having average roughness (*Ra*) of about 15.4 nm (as reported for AFM at 100°C). It consists of a scan size of 1×1 μm^2 and truncated cone surface structures (indicating self-similarity). A random distribution of structures is chosen to mimic the sputtering deposition used in this study. The fractal characterization functions of the model surface are shown in Figure 5.11 and are related to the experimental fractal profiles as follows:

   i. As shown, *A(r)* decreases monotonically to zero and then exhibits oscillatory characteristics about zero with the increasing values of shift (r). The oscillations are seen to increase with the increase in r.

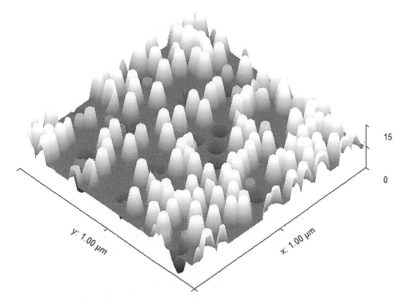

FIGURE 5.10 Three-dimensional model of mounded surfaces with regularly distributed and truncated cone-shaped structures.

ii. For the log-log profiles of the height–height correlation function ($H(r)$), the function increases linearly at small values r and then nearly stabilizes with oscillatory characteristics observed at large values of r.

iii. The double log profile of **PSDF** versus spatial frequency (K) of model surface exhibits two clear regions namely, the knee (or white noise or constant PSD) at low spatial frequencies and highly correlated (exponential decay) regions.

iv. The Minkowski connectivity curve for the model surface shows a slight decrease between z=0 and z=7.5 nm and then increases gradually. The profile is dominated by positive values as expected (since the generated model was dominated by valleys and peaks). This observation confirms that the Al thin films analyzed in this case exhibited fewer pits and pores. There were less pits and pores at 100°C than at the rest of the temperatures.

ESEM/XRD

The FESEM and XRD results of these surfaces were undertaken to validate the topological analysis of AFM data and as represented as Figures 5.12 and 5.13, respectively. The FESEM results in Figure 5.12 show that as the substrate temperature increases, the film becomes dense and continuous due to the gradual formation of new and large structures at higher substrate temperature [12]. At 40°C, the surfaces have segments of large, fine, and porous structures. We attribute the highest average values of D (Table 5.3) and the highest negative values of Minkowski connectivity to these observations. At 100°C, the microstructure is dominated by fine particles, agglomerated, and less porous structures which are

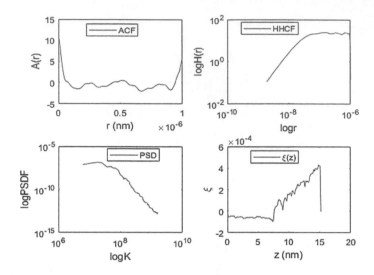

FIGURE 5.11 Autocorrelation function (ACF), height–height correlation function (HHCF), power spectral density (PSD) function, and Minkowski connectivity of mounded surface with truncated cone surface features shown in Figure 5.10.

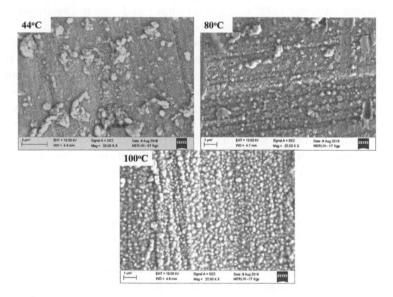

FIGURE 5.12 FESEM micrographs of aluminium thin films deposited on mild steel substrates at different substrate temperatures by RF magnetron sputtering.

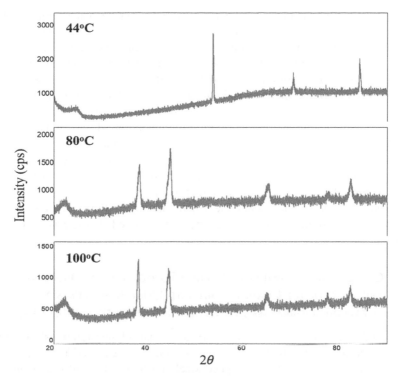

FIGURE 5.13 XRD patterns of the aluminium thin films deposited on mild steel substrates at different substrate temperature.

responsible for high roughness values (Table 5.1). Less porosity is associated with fewer pits on AFM images (Table 5.1) and this is the reason for very few negative values observed in the Minkowski connectivity curves in Figure 5.8. Further discussion on the effect of temperature of the structure of Al thin films has been reported in other publications [12].

The XRD patterns in Figure 5.13 indicate that the films are generally amorphous, and the number of crystalline peaks tends to increase with the substrate temperature. The presence of new peaks at higher temperatures indicates the growth of new and continuous surface structures. At 44°C, the peaks observed mostly measure the substrate due to the pits described through the AFM microscopy. These peaks disappear gradually with the increasing temperature since the surface coverage and connectivity of the Al structures increased at high temperatures.

Conclusion

We have used AFM technique to undertake a surface analysis of Al thin films deposited on mild steel substrates through RF magnetron sputtering at substrate temperatures of 44°C, 80°C, and 100°C. The different AFM images were subjected to statistical and fractal analysis to quantitatively study the surface structure. From the statistical analysis, we determined the interface width, average roughness, skewness, kurtosis, height distribution, maximum pit, and valley features. The fractal analyses were undertaken through autocorrelation, height–height correlation, and PSDFs. In addition, triangulation and cube counting techniques have been used to determine the fractal dimensions. The values of fractal dimensions obtained through the techniques are close indicating that all these techniques can be used to reliably approximate the fractal dimension of Al thin films. To further describe the patterning of the structural units of the fractal nature, 2D MFs were used. The results suggest that there is a significant influence of varying low deposition temperatures (from room temperature to 100°C) during magnetron sputtering of Al thin films. We have further undertaken a numerical study on computer generated surfaces to understand the fractal nature of the films and we have observed the occurrence of both self-affine and mounded characteristics of the Al thin films.

5.3.2 Case 2: Fractal Nature of Surface Topography and Physical Properties of the Coatings Obtained Using Magnetron Sputtering by Kwasny, Dobrzanski, Pawlyta, and Guilbinski (2004) [43]

In this study, the authors presented a very good and reliable relationship between fractal dimensions and statistical roughness characteristics of Ti+Ti (C, N) prepared via magnetron sputtering on high speed steel substrates. The substrates are composed of 8.5%wt. Co, 5.0%wt. Mo besides the other common constituents. Prior to sputtering, the substrates were heat treated followed by triple tempering. The sputtering process was undertaken at a vacuum pressure of 7×10^{-2} Pa in a neutral atmosphere composed of 50% CH_4 and 50% N_2. The deposition was undertaken for 1 hour and at increasing deposition temperature of 460°C, 500°C, and 540°C. The magnetron voltages used were 380 V, 420 V, and 395 V with a current varying between 5 and 6 amperes. The working pressure of deposition was maintained at 5.7×10^{-1} Pa.

The statistical and fractal analyses of the AFM data were then validated by the FESEM and XRD imaging and the results are physically significant to the magnetron sputtering process. The deposited thin films were evaluated for phase composition using Dron 2.0 XRD, texture properties using XRD7, microhardness using an ultra-microhardness tester, surface roughness using a Taylor–Hobson Sutronic device, and topography using AFM. The AFM measurements were undertaken at four different scans of 1, 2, 5, and 10 μm. The fractal analysis was undertaken to determine the fractal dimensions. This was done through the projective covering method (PCM). The method uses the area of squares of known scale and heights of the four corners to compute the fractal dimensions of the surfaces. A detailed theory of PCM for fractal dimension measurements are presented in the literature [44]. In this case, the AFM images, each constituting 512×512 pixels. Following the PCM, the areas of the respective subdivisions were determined and the results were used to plot on a bi-logarithmic scale. A least square method was used to fit the plot and the fractal dimensions were determined from the slope of the plots.

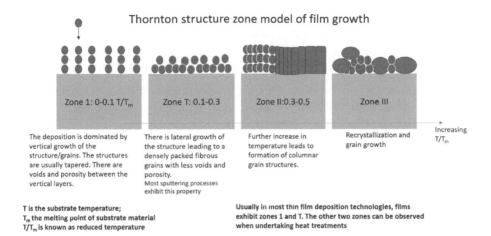

Thornton structure zone model of film growth

Zone 1: 0-0.1 T/T_m	Zone T: 0.1-0.3	Zone II:0.3-0.5	Zone III

Increasing T/T_m

The deposition is dominated by vertical growth of the structure/grains. The structures are usually tapered. There are voids and porosity between the vertical layers.

There is lateral growth of the structure leading to a densely packed fibrous grains with less voids and porosity.
Most sputtering processes exhibit this property

Further increase in temperature leads to formation of columnar grain structures.

Recrystallization and grain growth

T is the substrate temperature;
T_m the melting point of substrate material
T/T_m is known as reduced temperature

Usually in most thin film deposition technologies, films exhibit zones 1 and T. The other two zones can be observed when undertaking heat treatments

FIGURE 5.14   Schematic representation of Thornton's structure zone model of thin film growth.

The surface characterization of the films through scanning electron microscopy revealed that the films exhibited an amorphous structure and were evenly distributed. It was also observed that the films strongly adhered to the high-speed steel substrates. From these microstructural observations, the authors alluded the behavior to Zone II of Thornton's structural model. The Thornton's structural model is shown in Figure 5.14. The readers are referred to the literature [41,45–48] on structure zone models (SZMs) describing the growth of thin films during deposition. As shown, in zone II, the thin films are characterized by columnar grain structures. The roughness measurements in terms of the Ra (average roughness) statistical parameter showed that the highest values were obtained at the lowest process temperature (460°C) whereas the lowest roughness was obtained at 500°C. The thickness of the films was shown lowest at the lowest process temperature and highest at the temperature of the lowest roughness (500°C).

The fractal dimensions in this study were determined through the weighted average value of the fractal dimension values determined for the four ranges of scan size. This approach was employed because it has been stated that in the literature that the fractal dimension does not depend on the scan size as it is the case of statistical roughness values. For example, Mwema, Akinlabi, and Oladijo (2020) recently published a conference article in *Materials Today: Proceedings*, and presented a dataset illustrating the fractal the relationship between scan size (1, 3, and 30 μm) and fractal dimensions. From their data, it was observed that the fractal dimension does not significantly change with the AFM scan size. From other literatures [16,33], the scan size could be related to the magnification and it has been reported that magnification does not influence the fractal dimensions of the AFM microscopy. The highest fractal dimensions in this case study [43] were obtained at 460°C and 540°C. The highest values of fractal dimensions corresponded to the highest statistical roughness which indicated that the fractal dimension is an indicator of the irregularity and complexity of the thin film surfaces. In this study, therefore it can be concluded that statistical surface roughness exhibits a direct correlation with the fractal dimensions.

5.3.3 CASE 3: DEPENDENCE OF FRACTAL CHARACTERISTICS ON THE SCAN SIZE OF AFM PHASE IMAGING OF ALUMINIUM THIN FILMS, BY MWEMA, AKINLABI, AND OLADIJO (2020) [49]

In this case study, the authors of this book evaluated the relationship between scan size of AFM of phase imaging of aluminium thin films deposited via RF magnetron sputtering process. The aluminium thin films were deposited on stainless steel substrates at RF power of 150 W for 2 hours. The substrates were ground to fine surface finish to remove oxides and then they were cleaned with acetone. The samples were mounted on the sputtering system at a separation distance of 30 cm from the target. Prior to deposition, the chamber was vacuumed to pressures in the order of 10^{-5} Pa and then pre-sputtering was undertaken for 30 seconds to clean the Al target before the actual deposition. The deposition of Al thin films was undertaken at room temperature and the sample rotation speed of 5 revolutions per minute. After deposition, the films were left to cool inside the chamber for 8 hours to avoid contamination or oxidation of the surfaces of the Al thin films.

The samples were then sliced via a water jet in sizes of 10 mm × 10 mm for AFM measurement. The AFM measurements were undertaken in the following conditions: Integral grain of 0.6, the proportional grain of 0.8, setpoint of 0.8554, the temperature was maintained at room temperature, the scan rate of 1 Hz, and in the tapping mode. For each measurement, image sizes of 256 × 256 pixels were obtained and at three different scan sizes of 1 × 1 μm, 3 × 3 μm, 30 × 30 μm. The images were obtained using the phase channel of the AFM. For each surface, the AFM measurements were taken on 10 random regions for statistical accuracy. The obtained AFM images were characterized for statistical roughness using average surface roughness values, root means square roughness, skewness, kurtosis, etc. The fractal characteristics were determined for each measurement through the computation of the fractal dimensions using PSDF, box-counting method, triangulation, and partitioning method. It was noted that fractal dimensions obtained through the different methods were not significantly different and as such, the reported fractal dimensions in this paper were computed from the average values obtained from each method. The results on vertical and lateral roughness revealed the following:

i. The mean average surface roughness (Ra) increases with the scan size of the AFM imaging. The observation can be attributed to more features detected by the AFM probe when it travels at a larger scan scale.

ii. The values of the root mean square roughness (Rq) increase with the scan size of the AFM phase imaging. Additionally, at every scan size, the Rq values were larger than the Ra values.

iii. The average fractal dimensions (D) values did not vary significantly with the increase in the AFM scan size (magnification). However, it is important to note that the D values slightly increased with the scan size during phase imaging. This means that as the vertical irregularity and complexity of the thin films evolve, there is no considerable evolution of the lateral roughness with the scan size.

5.3.4 CASE 4: EFFECT OF ANGLE OF DEPOSITION ON THE FRACTAL PROPERTIES OF ZNO THIN FILM SURFACE BY YADAV ET AL. (2017) [20]

In this study, ZnO thin films of 150 nm thickness were sputtered on silicon substrates at room temperature. The aim of the research was to investigate the different angles of deposition to the fractal characteristics of the ZnO thin films. As such, the films were deposited at various angles namely, 20°, 30°, 40°, 60°, and 75°. The properties of the thin films were then studied via glancing angle XRD and AFM.

The AFM images were subjected to the computational procedure to determine the surface roughness and fractal characteristic parameters. The surface roughness was described using average roughness (Ra) and two-dimensional root mean square (Rq) whereas the fractal properties were described by fractal dimensions and Hurst exponents. The Brownian motion of the surface profile of the AFM images was used to examine the self-affinity of the roughness characteristics and Hurst exponents. The Higuchi algorithm was used to determine the fractal dimensions of the AFM micrographs.

The obtained results on surface roughness and fractal characteristics revealed that

i. the lowest fractal dimension was obtained at the second lowest average roughness and root mean square values,

ii. the second lowest fractal dimension was obtained at the highest average roughness and root mean square values,

iii. the highest fractal dimension was observed at the third lowest average surface roughness, the lowest Hurst exponent was also observed at this same average surface roughness.

5.3.5 RELATIONSHIP BETWEEN FRACTAL ROUGHNESS AND VERTICAL ROUGHNESS

From the preceding discussion, the relationship between the fractal and statistical (vertical) was presented. The following conclusions on the relationships were obtained.

i. In most cases (based on the existing literature) vertical roughness and lateral roughness do not have a direct correlation. This is mostly reported in complex structures of thin films such as oxides and alloys films.

ii. In most literature involving pure metallic thin films, there exist direct relationships between the surface roughness parameters and fractal properties. For instance, the fractal characteristics of Al thin films deposited through magnetron sputtering revealed that the lowest fractal dimension occurs at the highest roughness parameters [12]. In another study, it was reported that the highest fractal dimensions occurred at the highest surface roughness [50]. In a recent review, Mwema et al. [51] revealed that thin films of Indium-doped zinc oxide (IZO) deposited through sol-gel spin coating as reported by Ghosh and Pandey [52] exhibited a linear correlation between fractal dimension and root mean square roughness. It, therefore, means that this conclusion does not apply to all deposition processes. It would be an interesting area of research in the future.

iii. It has been idealized that during sputtering, the target material first hits and forms the films on the substrate surface region closer to the target [53]. Then, on continuous sputtering, the target material gets deposited to the free sites farther away from the target position. As such, in the sputtering of thin films of pure metallic materials, the films grow vertically and then laterally. Figure 5.15 illustrates the ideal evolution of surface structures during the sputtering of pure metallic films. At the initial stages, Figure 5.15(a), there is vertical stacking of the target atoms onto the surface of the substrates and the phenomenon in this region exhibits high Ra parameters. The vertical evolution of the process coincides with zones I, T, and II in the Thornton's SZM as shown earlier in Figure 5.14. In Figure 5.15(b), the continuous rotation of the substrate and influence of other underlying process parameters (such as temperature) results in the deposition of the target material to the entire surface of the substrate, resulting in the lateral development of the films on the surface of the substrate. The phenomenon in Figure 5.15(b) corresponds to zone III of the Thornton's SZM. The behavior leading to fractal evolution of thin films during sputtering and other deposition methods can be related to the capillary actions and surface tension theories of nucleation and grain growth described in the literature [54].

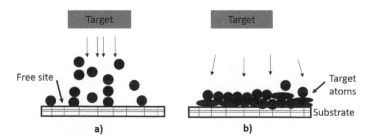

FIGURE 5.15 Illustrating the (a) vertical and (b) lateral development of pure metallic thin films during sputtering deposition.

5.4 INFLUENCE OF SPUTTERING PARAMETERS ON FRACTAL CHARACTERISTICS OF THIN FILMS

In this section, the influence of sputtering parameters on the fractal characteristics of thin films is presented based on some of the studies by the authors of this book and other literatures. The subsection is organized in terms of the sputtering parameters and their influence to the fractal behavior of thin films. According to the experience of the authors of this book, substrate temperature, power, and substrate type are key parameters that have been found to influence the fractal nature of sputtered thin films.

5.4.1 EFFECT OF SUBSTRATE TYPE

In one of the studies by Mwema, Akinlabi, and Oladijo in 2020, a detailed analyses of aluminium thin films sputtered on different substrates were presented [55]. The films were deposited in four metallic substrates namely Ti6Al4V, stainless steel, mild steel, and commercially pure titanium metal. The sputtering was undertaken at RF power of 150 W and for 2 hours. A high-purity Al target was used to deposit thin films of the thickness of 500 nm. All the substrates were prepared in the same procedure; grinding using silicon carbide (SiC) papers in the orders of #320, #500, #800, #1200, and diamond paste finishing of up to 1 μm. The substrate roughness values (Ra) ranged between 2 and 4 μm. An equal pre-sputtering of 30 seconds of all the substrates was undertaken before the deposition of the Al thin films. For each substrate, three samples were prepared to enhance the statistical accuracy of the measurements.

The samples were then characterized for surface topography using AFM at scan sizes of 3 × 3 μm. The obtained images were then taken through image analysis to determine the following:

  i. Interface width/root mean square roughness, Rq
  ii. Average roughness, Ra
  iii. Skewness and kurtosis
  iv. Morphological characteristics
  v. Height–height correlation functions
  vi. PSDFs
  vii. Minkowski functionals

In terms of statistical roughness parameters, there were considerable variations across the different substrates. The highest roughness values were obtained on the Al films deposited on mild steel substrates, followed by Ti6Al4V and pure Ti metal, whereas the lowest roughness values were observed on the stainless substrates. Similar trends were observed for the morphological characteristics. For instance, the maximum roughness valley depth and height of the roughness (for further description of these parameters, the reader is referred to the article by Gadelmawla [56]) were seen highest on mild steel and lowest on stainless steel substrates.

The fractal analyses via correlation and PSDFs revealed significant changes in the fractal characteristics of Al thin films on different substrates. It was shown that the highest fractal dimension values occurred on the films sputtered on pure titanium metal whereas the lowest values were obtained on films deposited on Ti6Al4V substrates. Based on the height–height correlation and PSDFs plots, all the substrates exhibited self-affine characteristics. The two-dimensional MFs were described by volume, boundary length, and Euler characteristic parameter (connectivity). The largest values of Minskowski volume were obtained on stainless steel substrates while the lowest values were observed on mild steel substrates. These results indicate that Al thin films grown on stainless steel exhibits less trenches, porosity, and pinholes contrary to the mild steel substrates. Additionally, the boundary length indicated that Al films deposited on stainless steel and Ti6Al4V exhibited similar fractal nature. For all the samples, it was shown that the Euler characteristic values were all positive and the highest value was obtained on Al films grown on stainless steel substrates. This means that stainless steel films exhibited less defects and the best lateral substrate coverage over all the other cases.

In similar research, it was demonstrated that the fractal characteristics of Al thin films deposited on mild steel and stainless steel substrates at similar sputtering conditions exhibited significant differences in fractal behavior [10]. It was also reported through a PSD analyses that Al thin films were grown on pure titanium and Ti6Al4V thin films exhibit different fractal characteristics [57]. These results indicate that the substrate type plays an important role in the growth of thin films. The results can be affirmed by the recent work of the authors of this book [58] and those by Khachatryan et al. [13] where it was demonstrated that Al films grown on glass exhibited small, well-defined, and low roughness as compared to those deposited on steel substrates. It was described that the surface chemistry and nature of substrate influence the diffusion, adhesion, nucleation, and growth of the Al films. Generally, metallic substrates exhibit higher roughness values and the structural evolution can be enhanced by the natural defective sites available on such substrates.

5.4.2 Effect of Substrate Temperature

The effect of substrate temperature on the sputtering process of thin films has been discussed widely and was also explained in Chapter 1 of this book. The SZM models of thin film deposition are developed based on substrate temperature [41]. In this subsection, the effect of substrate temperature to the fractal behavior of thin films is presented based on case studies from the authors' research and published literature.

An investigation on the effect of substrate temperature on Al thin films sputtered via RF magnetron sputtering on steel substrates was presented by Mwema et al. [59]. A PSD analysis was used to study the fractal characteristics of the AFM of the films deposited at temperatures of 40°C, 60°C, 80°C, and 100°C. The following results were obtained as it relates to temperature and fractal properties.

i. The PSD bi-log plots for all the samples within this temperature range exhibited similar behavior, that is, white noise region (at the low spatial frequency) and power law behavior at high spatial frequency. These plots were indications that all the samples were self-affine and hence fractal surfaces.

ii. The white noise regions were generally small for all the samples, which indicated laterally non-uniform characteristics of the films.

iii. The power density decreased with an increase in substrate temperature between 40°C and 100°C, which means that the lateral roughness increased with the substrate temperature.

iv. The correlation lengths were determined by fitting the data to K-model and it was shown that correlation lengths decreased with the temperature. This means that increase in temperature results in laterally non-uniform thin films. These results can be related to the influence of the substrate temperature onto the diffusion, nucleation, and grain growth (during film formation) of the thin films.

In a related study, Mwema et al. [53] deposited Al thin films on stainless steels at substrate temperatures of 44.5°C, 60°C, 80°C, and 100°C. Fractal characteristics were studied through fractal dimension computations and MFs. The fractal dimensions were computed using partitioning, cube counting, triangulation, and power spectrum and their average values obtained. It was shown that the fractal dimensions generally increased with the substrate temperature. This means that increasing the substrate temperature enhanced the growth and complexity of the surface structures. It was noted in this study that increasing the substrate temperature there is shrinkage of the columnar structures and lateral formation of structures leading to the formation of thin films as described earlier using the Thornton's model. These processes are responsible for the increase in fractal dimensions with the substrate temperature. The values of MFs were computed for all the samples and it was found that the values were negative and very small (in the order of 10^{-5}). These results imply that the pits and ripples were highly connected and the films exhibited a bi-continuous morphology. It was reported that an increase in substrate temperature leads to a decrease in the values of Minkowski connectivity (Euler characteristic value). This confirms that increasing the substrate temperature enhances the lateral development of thin films during a sputtering process.

In another study, Mwema et al. [60] studied the fractal characteristics of Al thin films deposited on glass substrates at 55°C, 65°C, and 95°C. The fractal dimensions were computed in a similar procedure as the study [53]. Contrary to other studies, the fractal dimension of the glass substrate decreased considerably from 3 upon sputtering (It is known that glass is fractal). The average fractal dimension was shown to decrease with an increase in substrate temperature. This can be attributed to the fact that introducing thin films of Al on fractal glass surface lowers the fractal nature of the glass substrate. Since an increase in the substrate temperature increases the formation rate of Al thin film structures (through enhanced surface mobility of adatoms and grain growth), the fractal effect of the glass is expected to reduce further. The comparative results of glass and steel substrates are further confirmation of substrate type influence on the fractal behavior of sputtered thin films. Using electron beam evaporation, Reza et al.[61] reported on the deposition of manganese thin films at increasing substrate temperatures (473, 623, 773, and 923 K). It was shown that fractal dimensions decreased with the increase in the substrate temperature of the Mn films during electron beam evaporation. It can therefore be stated that the relationship between the fractal properties and substrate temperature depends on the specific techniques of deposition and that the results presented here on sputtering processes cannot be inferred, without proof, to other related technologies such as thermal spray, electron beam, and chemical vapor deposition processes.

5.4.3 Effect of Sputtering Power

The sputtering power plays an important role in the generation of sputtering plasma and sustenance of the target-substrate processes. The authors of this book have published several articles on the influence of RF power of a magnetron sputtering system [62–65] In those publications, microstructure and mechanical properties of the Al thin films at different sputtering power were discussed. In the study [65], Mwema et al. investigated the influence of the RF power on the fractal behavior of Al thin films sputtered stainless steel substrates. The films were prepared at increasing powers of 150, 200, 250, and 350 W. The fractal analysis was undertaken on the (FESEM images and the fractal dimensions were shown to decrease with the increase in RF power. In the same study, the vertical roughness was computed from optical surface profilometry and the values did not exhibit any relationship with the RF power. It can therefore be deduced that increasing the RF power enhances spatial development with less complexity. This means that there occurs that specific RF power at which the best lateral evolution of the films can occur. A PSD analysis was used to determine the fractal behavior of Al thin films deposited on steel substrates [66] at RF power of 200 and 300 W. It was shown that the films deposited at 300 W were denser with less voids than the 200 W sample. These observations can be attributed to lower fractal dimensions obtained at higher RF power during the sputtering process. A resource of an overview of fractal analysis of thin film surfaces can be found in the study [67] from which most of the reported results can affirm these conclusions.

5.4.4 Other Parameters

In describing factors influencing the fractal behavior of sputtered thin films the reader is referred to other studies, which discussed the influence of other parameters besides temperature, substrate, and RF power. Reza et al. [68] has discussed the influence of argon gas flow rate and working pressure on the fractal behavior of ZnS. The effect of the angle of sputtering on the fractal characteristics of ZnO thin has been reported by Yadav et al. [20]. Hosseinpanahi et al. [69] investigated the influence of sputtering time on the fractal features of cadmium telluride (CdTe) thin films. In all these studies, the fractal features such as fractal dimension, and Hurst exponents exhibit different trends with the different parameters.

5.5 FRACTAL AND ROUGHNESS STUDIES IN MULTI-LAYER SPUTTERED THIN FILMS

The fractal characteristics of the multilayered thin film grown via the sputtering process should be interesting due to the nature of these films. Generally, these films consist of layers of varying properties across their interfaces, which implies that the fractal behavior of such films should be a composite of each of the constituting layers. The authors of the book have not been involved in the sputtering of multilayered thin films and therefore may not provide very detailed insights into this subject. However, based on their knowledge from literature and inference from single-layer thin films, important aspects of the fractal nature of such films are presented.

A study by Talu, Ghaderi, Stepien, and Mwema [70] reported on the fractal characterization of the bi-layer thin films of Cu/Fe nanoparticles deposited via sputtering. In the study, two sets of films were prepared – group I consisting of Cu 55 nm/ Fe 55 nm and group II consisting of Cu 55 nm/Fe 70 nm. The preparation was undertaken at similar conditions of sputtering such that the effect of multilayer thickness was investigated. It was shown that the group I samples exhibited the lowest fractal dimension and root means square (Rq) roughness values. The group II samples (with varying thicknesses of the bi-layers) exhibited the highest fractal dimensions and root means square roughness values. These results imply that the multilayering thickness variation can be detected through fractal theory. In a similar study, the height-to-height correlation was used to investigate the fractal characteristics of Ni/Ti multilayer films deposited via magnetron sputtering [71]. A power law correlation was observed to predict the relationship between the interface width and the bilayer thickness. The fractal evolution on multilayer systems was attributed to the argument that every time a new layer is deposited, the growth of the films 'restarts'.

It is noted that there is very scarce literature on fractal characteristics of sputtered multilayer thin films and therefore the subject is not well understood.

5.6 SUMMARY

In this chapter, mono-fractal analyses of the roughness of sputtered thin films have been discussed. Based on the presented case studies, mono-fractal analyses are important in evaluating the self-similarity or self-affinity in the sputtered thin films. Besides, the approaches can be used to further understand the evolution of the surface topography during the sputtering process. In thin pure aluminium films, the fractal dimensions and statistical roughness parameters (Ra and Rq) were shown to exhibit a direct correlation as the evolution of the surface features occurs. In alloys and complex materials, there was no trend reported in the relationship between fractal characteristics and statistical surface roughness parameters. Similar to the statistical roughness, fractal properties (lateral roughness) were shown to depend on the deposition conditions of the sputtering process, majorly, the type of the substrate, the sputtering power, and the substrate temperature. Other parameters reported in the literature in relation to the fractal evolution of sputtered thin films include time, thickness, and angle of deposition.

REFERENCES

[1] Ş. Ţălu et al., "Micromorphology analysis of specific 3-D surface texture of silver chiral nanoflower sculptured structures," *J. Ind. Eng. Chem.*, vol. 43, pp. 164–169, 2016.

[2] Ş. Ţălu et al., "Influence of annealing process on surface micromorphology of carbon–nickel composite thin films," *Opt. Quantum Electron.*, vol. 49, no. 204, 2017, pp. 1–9.

[3] J. Proost, M. Baklanov, R. Verbeeck, and K. Maex, "Morphology of corrosion pits in aluminum thin film metallizations," *J. Solid State Electrochem.*, vol. 2, no. 3, pp. 150–155, 1998.

[4] S. Srivastav, S. Dhillon, R. Kumar, and R. Kant, "Experimental validation of roughness power spectrum-based theory of anomalous cottrell response," *J. Phys. Chem. C*, vol. 117, no. 17, pp. 8594–8603, 2013.

[5] J. Li, Q. Du, and C. Sun, "An improved box-counting method for image fractal dimension estimation," *Pattern Recognit.*, vol. 42, no. 11, pp. 2460–2469, Nov. 2009.

[6] Ş. Ţәlu et al., "Microstructure and tribological properties of FeNPs@a-C:H films by micromorphology analysis and fractal geometry," *Ind. Eng. Chem. Res.*, vol. 54, no. 33, pp. 8212–8218, 2015.

[7] C.-L. Tien, T.-W. Lin, K.-C. Yu, T.-Y. Tsai, and H.-F. Shih, "Evaluation of electrical, mechanical properties, and surface roughness of DC sputtering nickel-iron thin films," *IEEE Trans. Magn.*, vol. 50, no. 7, pp. 1–4, 2014.

[8] F. M. Mwema, O. P. Oladijo, and E. T. Akinlabi, "The use of power spectrum density for surface characterization of thin films," in *Photoenergy and Thin Film Materials*, X.-Y. Yang, Ed. Hoboken, NJ: John Wiley & Sons, Inc., 2019, pp. 379–411.

[9] F. M. Mwema, E. T. Akinlabi, and O. P. Oladijo, "Correction of artifacts and optimization of atomic force microscopy imaging," in *Title: Design, Development, and Optimization of Bio-Mechatronic Engineering Products*, K. Kumar and J. Paulo Davim, Eds. USA: IGI Global, 2019, pp. 158–179.

[10] F. M. Mwema, O. P. Oladijo, T. S. Sathiaraj, and E. T. Akinlabi, "Atomic force microscopy analysis of surface topography of pure thin aluminium films," *Mater. Res. Express*, vol. 5, no. 4, pp. 1–15, Apr. 2018.

[11] N. Muslim, Y. W. Soon, C. M. Lim, and N. Y. Voo, "Influence of sputtering power on properties of titanium thin films deposited by Rf magnetron sputtering," *ARPN J. Eng. Appl. Sci.*, vol. 10, no. 16, pp. 7184–7189, 2015.

[12] F. M. Mwema, E. T. Akinlabi, O. P. Oladijo, and J. D. Majumdar, "Effect of varying low substrate temperature on sputtered aluminium films," *Mater. Res. Express*, vol. 6, no. 5, p. 056404, Feb. 2019.

[13] H. Khachatryan, S. Lee, K.-B. Kim, H.-K. Kim, and M. Kim, "Al thin film: The effect of substrate type on Al film formation and morphology," *J. Phys. Chem. Solids*, vol. 122, no. May, pp. 109–117, Nov. 2018.

[14] H. Garbacz, P. Wieciński, B. Adamczyk-Cieślak, J. Mizera, and K. J. Kurzydłowski, "Studies of aluminium coatings deposited by vacuum evaporation and magnetron sputtering," *J. Microsc.*, vol. 237, no. 3, pp. 475–480, 2010.

[15] N. G. Semaltianos, "Thermally evaporated aluminium thin films," *Appl. Surf. Sci.*, vol. 183, no. 3–4, pp. 223–229, 2001.

[16] Y. Gong, S. T. Misture, P. Gao, and N. P. Mellott, "Surface roughness measurements using power spectrum density analysis with enhanced spatial correlation length," *J. Phys. Chem. C*, vol. 120, no. 39, pp. 22358–22364, 2016.

[17] T. Jänsch, J. Wallauer, and B. Roling, "Influence of electrode roughness on double layer formation in ionic liquids," *J. Phys. Chem. C*, vol. 119, no. 9, pp. 4620–4626, 2015.

[18] R. Buzio et al., "Self-affine properties of cluster-assembled carbon thin films," *Surf. Sci.*, vol. 444, no. 1–3, pp. L1–L6, Jan. 2000.

[19] M. Nasehnejad, G. Nabiyouni, and M. G. Shahraki, "Dynamic scaling study of nanostructured silver films," *J. Phys. D. Appl. Phys.*, vol. 50, no. 37, 2017.

[20] R. P. Yadav et al., "Effect of angle of deposition on the Fractal properties of ZnO thin film surface," *Appl. Surf. Sci.*, vol. 416, pp. 51–58, 2017.

[21] R. P. Yadav, S. Dwivedi, A. K. Mittal, M. Kumar, and A. C. Pandey, "Fractal and multifractal analysis of LiF thin film surface," *Appl. Surf. Sci.*, vol. 261, pp. 547–553, 2012.

[22] R. P. Yadav, M. Kumar, A. K. Mittal, S. Dwivedi, and A. C. Pandey, "On the scaling law analysis of nanodimensional LiF thin film surfaces," *Mater. Lett.*, vol. 126, no. July, pp. 123–125, 2014.

[23] D. Liu, W. Zhou, J. Wu, and T. Huang, "Fractal characterization of graphene oxide nanosheet," *Mater. Lett.*, vol. 220, pp. 40–43, 2018.

[24] S. Ţălu et al., "Micromorphology analysis of sputtered indium tin oxide fabricated with variable ambient combinations," *Mater. Lett.*, vol. 220, pp. 169–171, 2018.

[25] D. Nečas and P. Klapetek, "Gwyddion: An open-source software for SPM data analysis," *Cent. Eur. J. Phys.*, vol. 10, no. 1, pp. 181–188, 2012.

[26] M. Aqil, M. Azam, M. Aziz, and R. Latif, "Deposition and characterization of molybdenum thin film using direct current magnetron and atomic force microscopy," *J. Nanotechnol.*, vol. 2017, pp. 1–10, 2017.

[27] D. Nečas and P. Klapetek, "One-dimensional autocorrelation and power spectrum density functions of irregular regions," *Ultramicroscopy*, vol. 124, pp. 13–19, 2013.

[28] F. Valle, M. Brucale, S. Chiodini, E. Bystrenova, and C. Albonetti, "Nanoscale morphological analysis of soft matter aggregates with fractal dimension ranging from 1 to 3," *Micron*, vol. 100, no. May, pp. 60–72, 2017.

[29] F. M. Mwema, O. P. Oladijo, S. Sathiaraj, and E. T. Akinlabi, "Corrigendum: Atomic force microscopy analysis of surface topography of pure thin aluminum films (2018 Mater. Res. Express 5 046416)," *Mater. Res. Express*, Jul. 2019.

[30] F. Hosseinpanahi, D. Raoufi, K. Ranjbarghanei, B. Karimi, R. Babaei, and E. Hasani, "Fractal features of CdTe thin films grown by RF magnetron sputtering," *Appl. Surf. Sci.*, vol. 357, pp. 1843–1848, 2015.

[31] Ş. Ţălu, R. Pratap, O. Sik, D. Sobola, and R. Dallaev, "How topographical surface parameters are correlated with CdTe monocrystal surface oxidation," *Mater. Sci. Semicond. Process.*, vol. 85, no. June, pp. 15–23, 2018.

[32] R. Lopes and N. Betrouni, "Fractal and multifractal analysis: A review," *Med. Image Anal.*, vol. 13, no. 4, pp. 634–649, Aug. 2009.

[33] M. Senthilkumar, N. K. Sahoo, S. Thakur, and R. B. Tokas, "Characterization of microroughness parameters in gadolinium oxide thin films: A study based on extended power spectral density analyses," *Appl. Surf. Sci.*, vol. 252, no. 5, pp. 1608–1619, 2005.

[34] I. Levchenko, J. Fang, K. (Ken) Ostrikov, L. Lorello, and M. Keidar, "Morphological characterization of graphene flake networks using minkowski functionals," *Graphene*, vol. 05, no. 01, pp. 25–34, 2016.

[35] C. Douketis, Z. Wang, T. L. Haslett, and M. Moskovits, "Fractal character of cold-deposited silver films determined by low-temperature scanning tunneling microscopy," *Phys. Rev. B Condens. Matter*, vol. 51, no. 16, pp. 11022–11031, 1995.

[36] N. Naseri et al., "How morphological surface parameters are correlated with electrocatalytic performance of cobalt-based nanostructures," *J. Ind. Eng. Chem.*, vol. 57, pp. 97–103, 2018.

[37] Ş. Ţălu, "Characterization of surface roughness of unworn hydrogel contact lenses at a nanometric scale using methods of modern metrology," *Polym. Eng. Sci.*, vol. 47, pp. 2141–2150, 2013.

[38] Ş. Ţălu, M. Bramowicz, S. Kulesza, and S. Solaymani, "Topographic characterization of thin film field-effect transistors of 2,6-diphenyl anthracene (DPA) by fractal and AFM analysis," *Mater. Sci. Semicond. Process.*, vol. 79, no. February, pp. 144–152, 2018.

[39] N. Spyropoulos-Antonakakis et al., "Selective aggregation of PAMAM dendrimer nanocarriers and PAMAM/ZnPc nanodrugs on human atheromatous carotid tissues: A photodynamic therapy for atherosclerosis," *Nanoscale Res. Lett.*, vol. 10, no. 1, pp. 1–19, 2015.

[40] A. Grayeli Korpi et al., "Minkowski functional characterization and fractal analysis of surfaces of titanium nitride films," *Mater. Res. Express*, vol. 6, no. 8, p. 086463, Jun. 2019.

[41] F. M. Mwema, O. P. Oladijo, S. A. Akinlabi, and E. T. Akinlabi, "Properties of physically deposited thin aluminium film coatings: A review," *J. Alloys Compd.*, vol. 747, pp. 306–323, May 2018.

[42] D. Siniscalco, M. Edely, J. F. Bardeau, and N. Delorme, "Statistical analysis of mounded surfaces: Application to the evolution of ultrathin gold film morphology with deposition temperature," *Langmuir*, vol. 29, no. 2, pp. 717–726, 2013.

[43] W. Kwaśny, L. A. Dobrzański, M. Pawlyta, and W. Gulbiński, "Fractal nature of surface topography and physical properties of the coatings obtained using magnetron sputtering," *J. Mater. Process. Technol.*, vol. 157–158, pp. 188–193, Dec. 2004.

[44] H. W. Zhou and H. Xie, "Direct estimation of the fractal dimensions of a fracture surface of rock," *Surf. Rev. Lett.*, vol. 10, no. 05, pp. 751–762, Oct. 2003.

[45] E. Wallin, "*Alumina thin films: From computer calculations to cutting tools*," Linkoping University, Institute of Technology, 2008.

[46] I. Miccoli, R. Spampinato, F. Marzo, P. Prete, and N. Lovergine, "DC-magnetron sputtering of ZnO:Al films on (00.1)Al 2 O 3 substrates from slip-casting sintered ceramic targets," *Appl. Surf. Sci.*, vol. 313, pp. 418–423, 2014.

[47] J. E. Yehoda and R. Messier, "Are thin film physical structures fractals?," *Appl. Surf. Sci.*, vol. 22–23, no. PART 2, pp. 590–595, May 1985.

[48] I. Petrov, P. B. Barna, L. Hultman, and J. E. Greene, "Microstructural evolution during film growth," *J. Vac. Sci. Technol. A Vacuum, Surfaces, Film.*, vol. 21, no. 5, pp. S117–S128, Sep. 2003.

[49] F. M. Mwema, E. T. Akinlabi, and O. P. Oladijo, "Dependence of fractal characteristics on the scan size of atomic force microscopy (AFM) phase imaging of aluminum thin films," *Mater. Today Proc.*, vol. 26, pp. 1540–1545, 2020.

[50] F. M. Mwema, E. T. Akinlabi, and O. P. Oladijo, "Effect of scan rate on AFM imaging on 3D surface stereometrics of aluminum films," *Mater. Today Proc.*, vol. 18, pp. 2315–2321, 2019.

[51] F. M. Mwema, E. T. Akinlabi, O. P. Oladijo, O. S. Fatoba, S. A. Akinlabi, and S. Tălu, "Advances in manufacturing analysis: Fractal theory in modern manufacturing," in *Modern Manufacturing Processes*, First., K. Kumar and J. P. Davim, Eds. UK: Elsevier, 2020, pp. 13–39.

[52] K. Ghosh and R. K. Pandey, "Fractal and multifractal analysis of In-doped ZnO thin films deposited on glass, ITO, and silicon substrates," *Appl. Phys. A*, vol. 125, no. 2, p. 98, Feb. 2019.

[53] F. M. Mwema, E. T. Akinlabi, O. P. Oladijo, and J. D. Majumdar, "Effect of varying low substrate temperature on sputtered aluminium films," *Mater. Res. Express*, vol. 6, no. 5, p. 056404, Jan. 2019.

[54] H. Frey and H. R. Khan, *Handbook of Thin-Film Technology*. Berlin, Heidelberg: Springer Berlin Heidelberg, 2015.

[55] F. M. Mwema, E. T. Akinlabi, and O. P. Oladijo, "Effect of substrate type on the fractal characteristics of AFM images of sputtered aluminium thin films," *Mater. Sci.*, vol. 26, no. 1, pp. 49–57, Nov. 2019.

[56] E. S. Gadelmawla, M. M. Koura, T. M. A. Maksoud, I. M. Elewa, and H. H. Soliman, "Roughness parameters," *J. Mater. Process. Technol.*, vol. 123, no. 1, pp. 133–145, 2002.

[57] F. M. Mwema, O. P. Oladijo, and E. T. Akinlabi, "The use of power spectrum density for surface characterization of thin films," in *Photoenergy and Thin Film Materials*, X.-Y. Yang, Ed. Hoboken, NJ: John Wiley & Sons, Inc., 2019, pp. 379–411.

[58] F. M. Mwema, E. T. Akinlabi, and O. P. Oladijo, "Microstructure and surface profiling study on the influence of substrate type on sputtered aluminum thin films," *Mater. Today Proc.*, vol. 26, pp. 1496–1499, 2020.

[59] F. M. Mwema, O. P. Oladijo, and E. T. Akinlabi, "Effect of Substrate Temperature on Aluminium Thin Films Prepared byRF-Magnetron Sputtering," *Mater. Today Proc.*, vol. 5, no. 9, pp. 20464–20473, 2018.

[60] F. M. Mwema, E. T. Akinlabi, and O. P. Oladijo, "Micromorphology of sputtered aluminum thin films: A fractal analysis," *Mater. Today Proc.*, vol. 18, pp. 2430–2439, 2019.

[61] R. Shakoury et al., "Stereometric and scaling law analysis of surface morphology of stainless steel type AISI 304 coated with Mn: A conventional and fractal evaluation," *Mater. Res. Express*, vol. 6, no. 11, 2019.

[62] F. M. Mwema, E. T. Akinlabi, O. P. Oladijo, and S. Krishna, "Microstructure and scratch analysis of aluminium thin films sputtered at varying RF power on stainless steel substrates," *Cogent Eng.*, vol. 7, no. 1, pp. 1–12, 2020.

[63] F. M. Mwema, E. T. Akinlabi, and O. P. Oladijo, "*Influence of sputtering power on surface topography, microstructure and mechanical properties of aluminum thin films*," in *Proc. of the Eighth Intl. Conf. on Advances in Civil, Structural and Mechanical Engineering - CSM 2019*, Birmingham City, UK. 2019, pp. 5–9.

[64] R. Ramarajan et al., "Optimization of Zn 2 SnO 4 thin film by post oxidation of thermally evaporated alternate Sn and Zn metallic multi-layers," *Appl. Surf. Sci.*, vol. 449, pp. 68–76, 2018.

[65] F. M. Mwema, E. T. Akinlabi, and O. P. Oladijo, "Exploring the effect of rf power in sputtering of aluminum thin films-a microstructure analysis," *Proc. Int. Conf. Ind. Eng. Oper. Manag.*, pp. 745–750, 2019.

[66] F. M. Mwema, E. T. Akinlabi, and O. P. Oladijo, "Two-dimensional fast Fourier transform analysis of surface microstructures of thin aluminium films prepared by Radio-Frequency (RF) magnetron sputtering," *Adv. Mater. Sci. Eng. Lect. Notes Mech. Eng. Springer, Singapore*, pp. 239–249, 2019.

[67] F. M. Mwema, E. T. Akinlabi, and O. P. Oladijo, "Fractal analysis of thin films surfaces: A brief overview," in *Advances in Material Sciences and Engineering. Lecture Notes in Mechanical Engineering*. Mokhtar Awang, Seyed Sattar Emamian, and Farazila Yusof, Eds., Singapore: Springer, 2020, pp. 251–263.

[68] R. Shakoury et al., "Optical properties, microstructure, and multifractal analyses of ZnS thin films obtained by RF magnetron sputtering," *J. Mater. Sci. Mater. Electron.*, vol. 31, no. 7, pp. 5262–5273, Apr. 2020.

[69] F. Hosseinpanahi, D. Raoufi, K. Ranjbarghanei, B. Karimi, R. Babaei, and E. Hasani, "Fractal features of CdTe thin films grown by RF magnetron sputtering," *Appl. Surf. Sci.*, vol. 357, no. September, pp. 1843–1848, Dec. 2015.

[70] S. Tălu, A. Ghaderi, K. Stępień, and F. M. Mwema, "Advanced micromorphology analysis of Cu/Fe NPs thin films," *IOP Conf. Ser. Mater. Sci. Eng.*, vol. 611, no. 1, p. 012016, Oct. 2019.

[71] S. Maidul Haque, A. Biswas, D. Bhattacharya, R. B. Tokas, D. Bhattacharyya, and N. K. Sahoo, "Surface roughness and interface width scaling of magnetron sputter deposited Ni/Ti multilayers," *J. Appl. Phys.*, vol. 114, no. 10, p. 103508, Sep. 2013.

6 Multifractal Characterization of Structure Evolution with Sputtering Parameters of Thin Films

6.1 INTRODUCTION

In Chapter 5, the applications of mono-fractal techniques on thin film characterization were demonstrated. It was discussed that there are several methods of mono-fractal analyses including, power spectral density, triangulation, box counting, autocorrelation, and height–height correlation functions just to mention a few. It was also shown that box counting is the most suitable procedure for evaluation of mono-fractal characteristics of thin films and fractal dimension is the analytical index used to show the scaling behavior of the morphological features of thin films. However, in most structures, the scaling behavior (or the fractal dimension) varies from one region to another, and in such cases, the fractal dimension cannot fully describe the scaling behavior of the films. In such instances, the surfaces are said to exhibit multifractal behavior and the multifractal algorithm is used to describe the fractal characteristics of the surfaces. In a multifractal system, the fractal dimension does not sufficiently describe its dynamic fractal nature and in such cases, a continuous spectrum of exponents, usually known as the singularity spectrum is used. Some of the algorithms of multifractal behavior were described in Chapter 3 for the multifractal characterization of thin films and other phenomena.

There is a lot of literature describing multifractal characterization of surfaces although not limited to thin films. For instance, a two-dimensional multifractal detrended fluctuation analysis (MFDFA) was used by Liu et al. [1] to study the fractured surfaces of foamed polypropylene/polyethylene blends at different temperatures. It was found that the detrended fluctuation function and the scale exhibit a power law relationship. For such surfaces, it was also found that the scaling component has a nonlinear relationship to the moment order. The evolution of the temperature was also related to the multifractal spectra and it was shown that the surfaces were more irregular and complex as the content of polyethylene increased. These results implied that fractured surfaces of polypropylene/polyethylene blends exhibit multifractal characteristics.

In another study by the authors of this book, a mono-fractal and multifractal analyses were undertaken on Al thin films deposited on glass substrates at varying substrate temperature through radio-frequency (RF) magnetron sputtering. The mono-fractal characterization was undertaken on atomic force microscopy (AFM) images whereas the multifractal study was undertaken on field-emission scanning electron microscopy (FESEM) images. The FESEM images were taken through an image segmentation process and then the analyses were implemented in FracLac plug-in embedded in Fiji open source software. Based on the multifractal study, the following inferences were deduced from this study.

  i. The multifractal spectrum plots (multifractal spectra vs. singularity index) for all the samples were skewed to the right with a short hump to the left.
 ii. All the values of the width of the multifractal spectra were positive.
iii. The highest values of the singularity index were obtained at the highest substrate temperature which implies that the surfaces at high temperatures were more complex than those deposited at low substrate temperatures.
 iv. The generalized fractal dimension decreased gradually with moment order and then stagnated without further variation.
  v. These results indicated that the Al thin films deposited on glass substrates were mono-fractal.

Another illustrative study on applications of the multifractal theory was undertaken by Xu et al. [2] during the analysis of corrosion damage on steel reinforcing bars for construction application. Similar to the previous article, the researchers compared both mono-fractal and multifractal analysis of the corrosion damage. It was observed that corrosion morphologies of steel bars are either connected or isolated and exhibit irregular shapes and orientations. It was shown through mono-fractal analysis that the distribution of corrosion morphology exhibits a statistical fractal feature. In terms of the multifractal analyses, all the spectra were humped and mainly hooked to the left. All the values of widths of multifractal spectra were positive, which indicated that the distribution of corrosion depended on the maximum subset of probability and the generalized dimensions spectra were all decreasing with the moment order, which indicates that the corroded surfaces of the steels were multifractal.

The general procedure for undertaking a multifractal analysis involves the following steps for surfaces and it has been summarized in Figure 6.1.

  i. The first step is to obtain images of surface topography of the surfaces using scanning probe microscopy such as AFM, scanning tunneling microscopy (STM), and scanning probe electrochemistry (SPE), or surface microstructure using techniques such as scanning electron microscopy (SEM), optical microscopy, FESEM, transmission electron microscopy, and so forth.
 ii. The obtained images are then segmented, which involves partitioning the images into distinct regions. The regions are differentiated by pixels, such that features of the image with similar characteristics are assigned the same pixel. In such a case, the pixels exhibiting similar features such as

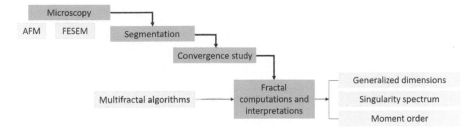

FIGURE 6.1 A summary of the multifractal characterization procedure of thin films adopted in this study.

color, intensity, or texture are clustered together and taken to represent a single feature of the image. In this process, the objective is to recognize features, their shape, and boundaries. There are various techniques of image segmentation and the readers are referred to the literature for detailed understanding [3,4].

iii. The segmented images are then analyzed for fractal characteristics through the following sequence steps. In general, the authors have experience in using box counting techniques to undertake multifractal characterizations and these steps are based on such a method. (a) Convergence study to determine the optimal number of grids for the analysis. It has been shown that the higher the number of grids (box counting), the accurate the analysis. However, very fine grids constitute to longer computation time and high cost of the computing facility. On the other hand, very rough grids result into unstable/inaccurate computations. As such, a trade-off between the cost/time and accuracy should be stricken through a convergence study, (b) based on the convergence study, the number of grids is input into the algorithm for computing the multifractal features of the image, and then finally (c) the actual computations and interpretation of multifractal parameters such as generalized dimension and singularity spectrum.

In this chapter, the multifractal characterizations of thin films grown via magnetron sputtering are described with examples of original and published data from the authors' works and literature, respectively. The studies presented are generally based on the procedure described in Figure 6.1 for the multifractal characterization of thin films. The emphasis is made on the relationship between multifractal characteristics and thin film growth mechanisms.

6.2 MULTIFRACTAL STUDIES OF SPUTTERED PURE METALLIC FILMS

In this topic, the multifractal studies on pure metallic films are presented. The authors of this book have published a lot of works on multifractal studies of aluminium thin films prepared by magnetron sputtering. Based on this expertise, detailed and step-by-step procedures for multifractal analyses of pure aluminium

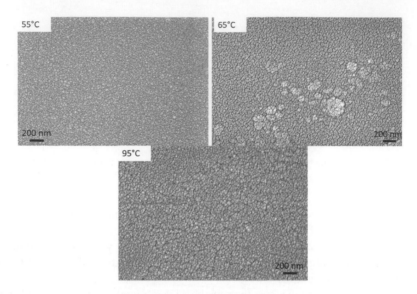

FIGURE 6.2 Field emission scanning electron microscopy images of aluminium thin films sputtered on glass substrates at different substrate temperatures. All the images were taken at the same magnification with a scale bar of 200 nm.

thin films deposited on metallic and glass substrates will be discussed here. It has been shown that fractal studies can be undertaken effectively on FESEM and AFM [5–8]. As a case study and for illustration of multifractal descriptions of thin films of pure metallic materials, herein presented is the multifractal characterization of pure aluminium thin films deposited on glass substrates at varying substrate temperatures of 55°C, 65°C, and 95°C.

Figure 6.2 shows the FESEM micrographs for pure aluminium thin films deposited on glass substrates. In brief, these films were deposited on glass substrates using RF magnetron sputtering system at constant power and argon gas flowrate but at three different substrate temperatures of 55°C, 65°C, and 95°C. The observations on microstructural evolutions were undertaken using FESEM and the results are shown in Figure 6.2. The discussions on microstructural evolution during the deposition were presented in an earlier publication and readers are referred to Mwema, Akinlabi, and Oladijo (2019) for further details on substrate preparation, deposition, and microstructural descriptions [8]. Here, the authors focus on illustrating the multifractal analyses and interpretation of the results for pure Al thin films based on the FESEM micrographs. While a lot of publications are available on these analyses, the step-by-step procedures are lacking in such publications since the focus is usually on the property-process relationship.

The first step on this procedure is the segmentation of the FESEM images. In this case, the thresholding method was used to segment the image to isolate the features required for the study. The features of interest were marked black while the background was marked white as shown in Figure 6.3 for thin films deposited at 55°C (for illustration purposes). The thresholding process should be carefully undertaken to ensure all the features of

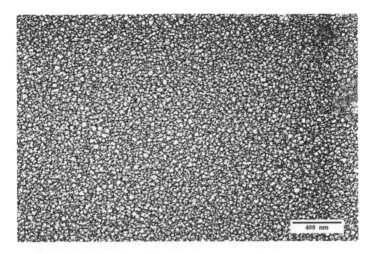

FIGURE 6.3   Segmented image (thresholding) of FESEM of Al thin films deposited on glass substrates at 55°C.

interest are identified. In this work, the process was implemented in the Fiji image processing tool of thresholding. The FESEM image was uploaded on Fiji software and it was then scaled based on the normal procedure (*Draw the line on the FESEM scale bar→ Menu→ Tools→ Analyse→ Set scale→ enter known distance→ click Global→ OK*) [9]. The images were then resized and taken through the thresholding process. During thresholding in Fiji software, careful attention should be taken to ensure the right features are separated from the image background. The procedure requires experience on the microstructure and one should know what they are analyzing. As shown in Figure 6.3, the features analyzed are indicated by dark color whereas the background is indicated by the white color.

After thresholding, the images are then uploaded into the FracLac plugin for ImageJ/Fiji for fractal analyses [10]. The FracLac was developed by Audrey Karperien of the Charlses Sturt University as a plugin to Fiji (ImageJ) software and it is used for the following applications.

   i. To measure complex geometrical shapes, which are otherwise difficult to measure using the known dimensional methods.
   ii. It finds applications in patterned structures such as those for biological forms and complex textures of naturally known fractal objects.
   iii. The plugin can be used to extract patterns from images and convert them into binary digital images for easy analysis.

The software undertakes different types of image analysis and the main outputs of the FracLac include

   i. fractal dimension measurements,
   ii. lacunarity analysis,
   iii. multifractal characterizations, and
   iv. size and shapes of patterns in binary/threshold images.

In this case, the multifractal measurements were based on the box counting technique. As such, the first step in the FracLac process was to undertake the convergence study to determine the minimum number of grids to provide a stable value of the fractal dimension. As shown in Figure 6.4, the lowest number of grids of the box counting method used were 12 in which the fractal dimension value of 1.83 was obtained. As it has been stated in the literature, the finer the grid (and the larger the number of grids), the accurate the box counting computation [11]. However, the larger the number of grids, the longer the computation time and the higher the cost of computing as powerful computing hardware would be required. As such, the optimal number of grids should be determined by undertaking a convergent study as shown in Figure 6.4 for Al thin films sputtered at 55°C and it can be seen that the lowest number of grids which gave the grid-non-independent fractal dimension was 800. These procedures were repeated for the other samples (65°C and 95°C) and then the results are presented in Figures 6.5 to 6.8.

Next, each of the multifractal measurements were undertaken using these number of grids (800). During multifractal analyses, the FracLac tool masks the threshold images to understand the background and foreground of the images to accurately extract the fractal characteristics of the necessary features (Figure 6.9). The first multifractal parameter computed in this method was the generalized multifractal dimension (D_q) spectra, which is usually provided as a plot of D_q versus moment order (q). The computation was undertaken at moment order values ranging between −10 and +10, which is usually the default range recommended for FracLac analysis and at increments of 0.25 [10]. The parameter computation is based on the equations described earlier in Chapter 3. The results of generalized dimension for all the three samples are shown in Figure 6.10. The D_q versus q multifractal generalized plot

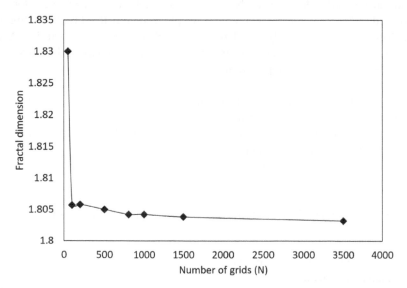

FIGURE 6.4   Convergence study to determine the optimal number of grids for the multifractal study of thin films deposited at 55°C on glass substrates.

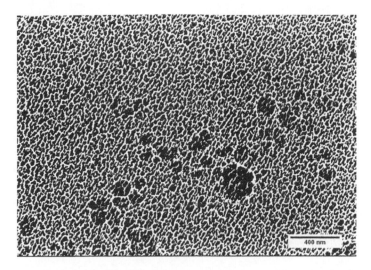

FIGURE 6.5 Segmented image (thresholding) of FESEM of Al thin films deposited on glass substrates at 65°C.

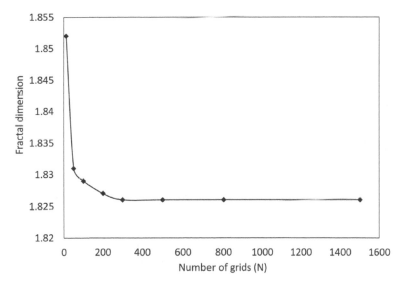

FIGURE 6.6 Convergence study to determine the optimal number of grids for the multifractal study of thin films deposited at 65°C on glass substrates.

indicated differences for the three deposition temperatures of the Al thin films and from these plots, the following results can be deduced:

i. The function D_q decreased with the increasing moment order q and exhibits sigmoidal behavior at q = 0. This is an indication that all the films exhibited a multifractal behavior. For mono-fractals and non-fractal images, this function (plot) exhibits a flat characteristic.

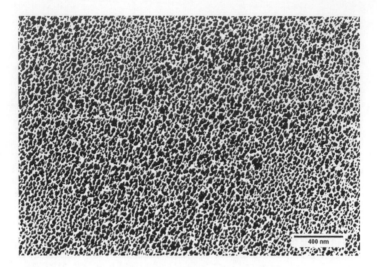

FIGURE 6.7 Segmented image (thresholding) of FESEM of Al thin films deposited on glass substrates at 95°C.

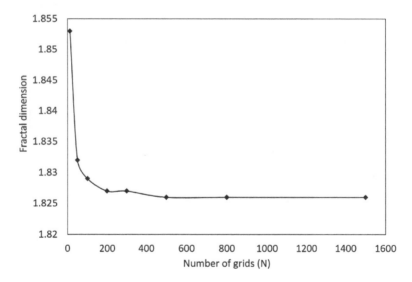

FIGURE 6.8 Convergence study to determine the optimal number of grids for the multifractal study of thin films deposited at 95°C on glass substrates.

ii. The most significant parameters for this multifractal behavior are $D_{(q=0)}$, $D_{(q=1)}$, and $D_{(q=2)}$. The values of the function at $q = 0$, results into a generalized dimension known as the *capacity dimension*. From Figure 6.9, the values for capacity dimensions were obtained as 1.829, 1.840, and 1.880 for films deposited at 55°C, 65°C, and 95°C, respectively. These values correspond to the box counting fractal dimensions of the thin films. The value of the function at $q = 1$ is known as *information dimension* and in this case, they were obtained as

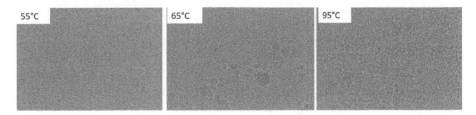

FIGURE 6.9   Illustrating the masking process of the FESEM images during multifractal analysis.

follows: 1.840 for 55°C, 1.839 for 65°C, and 1.877 for 95°C. The value of the function at q = 2 is known as the *correlation dimension* and here the following values were obtained for the three substrate temperatures; 1.847, 1.842, and 1.878, respectively. For multifractality, a general rule of D (0) > D (1) > D (2) is applied. However, in this case, the rule cannot be generalized which indicates the complex nature of the microstructure of Al thin film. As such, in some cases, the values of multifractal spectra, W, (D (0) − D (1)) may give negative values. These values are used to measure the extent of heterogeneity and the larger the value of W, the higher the non-uniform distribution of the morphological features in a thin film deposition process.

iii. The multifractal characteristics exhibited by the function is based on the literal definition of 'multifractals' as 'fractal structures which cannot be described by a single value of fractal dimension'. In this case, therefore, there are three general dimensions which can describe the fractal nature of the Al thin film deposited at different substrate temperatures. The details on the relationship between the D(q) function and the substrate temperature evolution can be further understood from published literature [8,12,13].

iv. Generally, the generalized fractal dimension parameter can be used to detect differences between thin films deposited at varying sputtering conditions and parameters. For instance, as shown in Figure 6.10, the profiles for the D(q) versus q shifts upward as the substrate temperature increases. This means that at a higher substrate temperature (95°C), the fractal dimensions describing the multifractality of the films are higher compared to those obtained for films at lower substrate temperatures. The conclusion is in agreement with the models describing the influence of temperature on the thin film sputtering processes described in the literature [14,15]. As the substrate temperature increases, the complexity of the microstructure development is expected and hence the spatial (fractal and lateral roughness) evolution of the surface structures would be expected at higher substrate temperatures.

The generalized moment ordering of the function is presented in Figure 6.11. As shown, at 55°C, the general fractal dimension increases gradually with the moment order and reaches a maximum after q = 1. For samples deposited at 65°C and 95°C, the dimension decreases gradually to q = 0.5 beyond which it stabilizes to a constant value of about 1.83. For samples deposited at a substrate temperature of 95°C, the

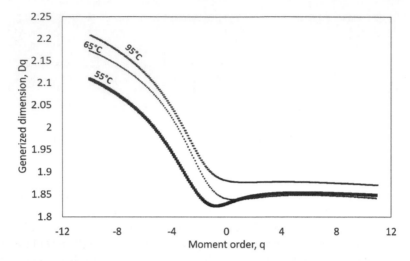

FIGURE 6.10   Generalized fractal dimension (D_q) versus moment order, q for the aluminium thin films deposited at 55°C, 65°C, and 95°C.

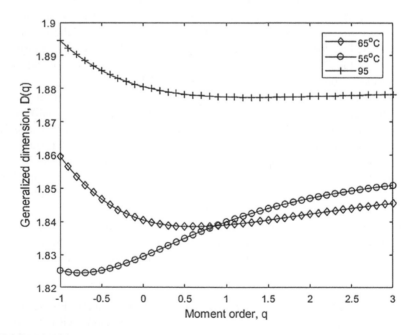

FIGURE 6.11   Moment ordering of the generalized fractal dimension for the films sputtered at different substrate temperature. [The moment ordering of the fractal dimension shows that higher stability is obtained on films exhibiting stronger multifractal behavior (95°C).]

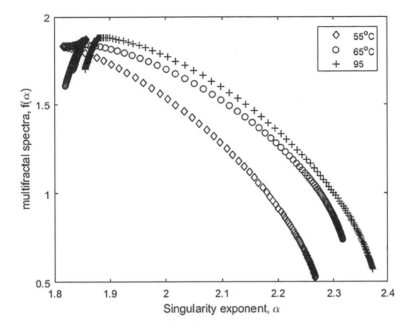

FIGURE 6.12 The multifractal spectra of aluminium thin films deposited on glass substrates at substrate temperatures of 55°C, 65°C, and 95°C. [All the spectra plots were skewed to the right with short hooks to the left. The films deposited at low temperatures exhibit the largest values of multifractal spectrum.]

generalized fractal dimension decreases and stabilizes to a constant value of 1.88 at $q = 0$. These results indicate that the moment ordering of the generalized fractal dimension is influenced by the substrate temperature. The moment ordering reveals different fractal dimensions indicating the multifractality of the films. It can further be seen that the samples exhibiting complex structures (95°C in this case) have their moment ordering profiles exhibiting larger fractal dimensions and therefore tend to be shifted upward compared to the rest of the surfaces.

Another important parameter for the multifractal characterization of surfaces is the multifractal spectrum function $f(\alpha)$, which is expressed against the singularity exponent (α). For this study, the multifractal spectra for the three samples prepared at different substrate temperatures are presented in Figure 6.12. The following are general results that can be deduced from these multifractal spectra plots.

i. The shape of the spectra $f(\alpha)$ versus the singularity exponent (α) determines whether the surfaces are non-fractal, mono-fractal, or multifractal. When the plot is humped regardless of the side of skewness, the scaling is said to be multifractal. If the plot converges at the top and the plot is skewed in one side, the surfaces are either non- or mono-fractal. As can be seen in Figure 6.12, the plots are humped with strong skewness to the right, which indicates that the surfaces are multifractal. However, the multifractality is not very strong as per the FracLac software descriptions [10].

ii. The width and peaks of the spectra plots are also used in describing the strength of the multifractal characteristics of surfaces. As shown in Figure 6.12, the peaks of the spectra shift with the increasing substrate temperature and the highest peak was observed on the samples obtained at 95°C. The α_{min} and α_{max} values which correspond to the least and largest singular strengths, respectively, are used to describe the width of spectrum functions. The corresponding multifractal functions are $f(\alpha_{min})$ and $f(\alpha_{max})$ and these are known as spectral values and correspond to fractal dimensions of the surfaces. From these values, the width of the multifractal spectrum $\Delta\alpha$ (computed as $\alpha_{max} - \alpha_{min}$) can be determined. The values for these parameters are shown in Table 6.1. It can be seen that the width of the spectrum function increased with the increasing temperature of the glass substrate. The spectral values ($\Delta f(\alpha)$) were all negative, implying that the distribution of surface structures depends on the minimum subset of probability. Additionally, the parameter $\Delta f(\alpha)$ can be used to statistically describe the ratio of the number of sites of the peaks and valleys. As such, the negative values indicate that the surfaces consisted of a greater number of highest peaks than the number of lowest/deepest valleys. The films deposited at maximum substrate temperature exhibited the largest value (-ve) indicating that it had less valleys, porosity, hillocks, and other defects. These results correlate with the hillock characterization reported for the same samples in an earlier publication by the authors of this book [16].

The multifractal scaling exponent $\tau(q)$ is another parameter used for multifractal characterization. The mass exponent ($\tau(q)$) was computed in this study at a moment order (q) in the range of -10 and 10 (Figure 6.13). It can be seen that the $\tau(q)$ exhibits a nonlinear relationship with q with a modulation point at $q = 0$. It can also be observed that the mass exponent plots of the films at different substrate temperatures exhibit a similar trend within the said moment order range.

In an earlier study by the authors of this book on these samples revealed some correlations between fractal characterizations and mechanical measurements (via nanoindentation techniques). Firstly, it was shown that the highest values of hardness were obtained on the samples deposited at 95°C whereas the lowest values on samples were prepared at 55°C. The mono-fractal measurements of 2D fractal dimensions based on AFM of the films showed that the highest fractal dimension was obtained at samples deposited at a substrate temperature of 55°C whereas the lowest value was obtained on films deposited at a substrate temperature of 95°C.

TABLE 6.1
Multifractal Spectra Parameters for Al thin Films Deposited on Glass at different Substrate Temperatures

Substrate Temperature (°C)	α_{min}	α_{max}	$\Delta\alpha = \alpha_{max} - \alpha_{min}$	$f(\alpha_{min})$	$f(\alpha_{max})$	$\Delta f(\alpha) = f(\alpha_{max}) - f(\alpha_{min})$
55	1.835	2.267	0.432	1.705	0.525	−1.18
65	1.82	2.316	0.496	1.605	0.74	−0.865
95	1.856	2.372	0.516	1.705	0.57	−1.135

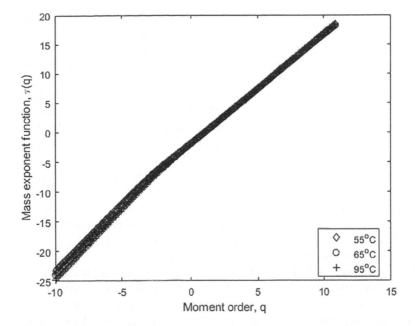

FIGURE 6.13 Plots of mass exponent $\tau(q)$ versus moment order (q) for the Al thin films deposited on glass substrates at different temperatures.

It was also shown that the highest multifractal spectra width is obtained on films deposited at the highest substrate temperature. In a similar study, Mwema, Akinlabi, and Oladijo (2020) conducted a micromorphology and nanomechanical characterization of pure Al thin films sputtered on stainless steel substrates at a varying substrate temperature of 44.5°C, 60°C, 80°C, and 100°C [17]. The work was published in *Materialwiss. Werkstofftech* in Wiley publishers and it was shown that as the substrate temperature increased the fractal characteristics of the pure Al thin films evolved [17]. The films deposited at temperatures between 44.5°C and 80°C were shown to exhibit mono-fractal behavior whereas those deposited at 100°C showed multifractal behavior. The study further undertook a relationship between the surface roughness and the fractal behavior of the films and it was seen that the width of the power spectrum exhibited an inverse relationship with the surface roughness. The surface roughness was shown to decrease with the increasing substrate temperature, whereas the width of the multifractal spectrum increased with the increasing substrate temperature of the Al thin films during sputtering. The largest values of spectrum function obtained at 100°C are an indication that those films are vertically irregular due to the influence of the substrate temperature to the sputtering process as it was demonstrated in another study by Mwema and co-authors [18,19]. The mono-fractal films were also shown to exhibit higher lateral roughness as compared to the multifractal thin films. This is an interesting finding and requires further investigations to determine the practical implications of the microstructural design of such films. Additionally, the study [17] revealed that films exhibiting mono-fractal characteristics have mass exponent functions nearly linear without a clear point of inflection contrary to the results represented in Figure 6.13. Another interesting finding to mention from that study [17] is that the Young's modulus and hardness were shown to be very high on thin films exhibiting the highest mono-fractal characteristics although these properties were seen to increase considerably with thin film surfaces tending to multifractality.

In one of the oldest studies on multifractal characterizations of thin films by Blacher et al. [20], the applications of the theory to polymer alloys and granular discontinuous metallic thin films were demystified. As was analyzed by the authors of this book in the study [17], there is a relationship between the multifractality and mechanical properties of thin films. Additionally, the multifractal theory can be used to detect the differences induced in microstructures of thin films by growth processes and conditions either to the process or to the substrates. The data on generalized fractal dimension versus the moment order can be useful in determining the non-uniformity of the thin films of pure metallic materials. It was reported that the greater the difference between $D_{(q=0)}$ and other values (say D at q = 1, 2, 3, …, n), the more the nonhomogeneous the microstructure of the films.

In another study [6], the effect of RF power on sputtered pure aluminium thin films was investigated by the authors of this book. The films were deposited on stainless steel substrates using the HHV TF500 physical vapor deposition facility. The details of the deposition facility have been described extensively by the publications of the authors [21,22]. The sputtering was undertaken at four different RF powers of 150, 200, 250, and 350 W for 2 hours and at a temperature of 90°C. The microstructural characterization and topography were obtained using FESEM and non-contact optical profilometer, respectively. The details of the FESEM and profilometry can be obtained in an earlier publication by the authors [23]. Then, the FESEM images were taken through image analysis for multifractal characterization and only multifractal spectra function was used (Figure 6.14). From the spectra, the multifractal parameters shown in Table 6.2 can be computed.

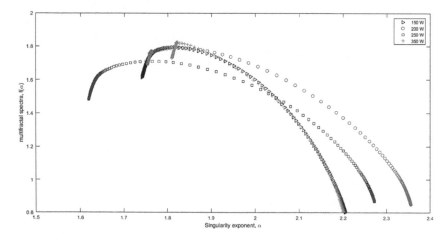

FIGURE 6.14 The multifractal spectra of aluminium thin films sputtered on stainless steel substrates at different RF power as shown (obtained from Mwema, Akinlabi, & Oladijo [6] under Attribution-Only, or CC-BY license).

TABLE 6.2

Multifractal Spectra Parameters for Al thin Films Sputtered on Stainless Steel Substrates at different RF Powers

RF Power (W)	α_{min}	α_{max}	$\Delta\alpha$	$f(\alpha_{min})$	$f(\alpha_{max})$	$\Delta f(\alpha)$
150	1.741	2.2	0.459	1.618	0.802	−0.816
200	1.754	2.36	0.606	1.7	0.852	−0.848
250	1.619	2.272	0.653	1.485	0.869	−0.616
350	1.81	2.202	0.392	1.732	0.805	−0.927

It can be seen from Table 6.2 and Figure 6.14 that the multifractal spectra width, $\Delta\alpha$, does not exhibit a direct correlation with the RF sputtering power of the aluminium thin films. As observed, the highest width was obtained at 250 W while the lowest value of spectra width occurred at 350 W. The spectral values, $\Delta f(\alpha)$, were all negative indicating that the spectra were humped and skewed to the right and exhibited fewer valleys than peaks. The largest absolute value of the spectrum was obtained at 350 W whereas the smallest at 250 W. The influence of RF power on microstructure and scratch characteristics of pure aluminium thin films was recently illustrated by Mwema et al. (2020)[24] and it was observed that the properties of films deposited within the range of 150 and 350 W through RF sputtering do not exhibit a correlation with the RF power. As such, the lack of multifractality-RF power trend (Figure 6.14 and Table 6.2) agree with such previous observations.

In another work by the authors titled, 'A Multifractal Study of Al Thin Films Prepared by RF Magnetron Sputtering' [25] published in *Lecture Notes in Mechanical Engineering* under the title, 'Advances in Manufacturing' Springer Singapore, a multifractal characterization of pure aluminium thin films deposited on mild steel and stainless-steel substrates at 150 W and 200 W was undertaken. The AFM was used to obtain the surface topography of the thin films and then the AFM images were taken through the image analysis procedure (FracLac Software) for multifractal studies. In this work, mass exponent function, multifractal spectrum as a function of singularity index, and generalized fractal dimension were used as measures of multifractality. The following key observations were made in this study.

i. The influence of the substrate temperature and RF power on the sputtering of pure aluminium thin films can be detected via multifractal characterizations. It was seen that the mass exponent function was not a very effective parameter for differentiating the effects of the two conditions. However, it was shown that singularity strength function and generalized fractal dimension parameters can be used to detect the differences.

ii. The multifractal singularity strength (multifractal spectrum) was seen to exhibit similar trends for different substrates in terms of the varying RF power. The spectral width ($\Delta\alpha$) values revealed that the films deposited on stainless steel substrates and 150 W were the least multifractal since these films had the smallest value of the spectral width. The samples prepared on mild steel and at 150 W on the other hand were seen to exhibit the largest value of spectral width and therefore they were more multifractal.

iii. The generalized fractal dimension, D(q), expressed as a function of the moment order, q, was also used to characterize the multifractality of the films in the said study. It was observed that all the films exhibited decreasing D(q) function with a point of inflection at q=0, indicating the multifractality of the films. However, it was noted that films grown on stainless steel substrates at 150 W exhibited a slower decrease of D(q) confirming that such films were less multifractal as compared to the rest of the films.

In another article titled, 'Effect of AFM Scan Size on the Scaling Law of Sputtered Aluminium Thin Films' by the authors was published by *Lecture Notes in Mechanical Engineering* under the same title as above, the influence of AFM measuring parameters on the multifractal parameters of thin aluminium films was illustrated [26]. The films were deposited on steel substrates at constant RF power of 150 W and then an AFM equipment was used to obtain the surface topography of the films. The AFM measurement was undertaken at different scan sizes (3 × 3 μm, 5 × 5 μm, and 10 × 10 μm). The images were then analyzed for multifractal behavior to illustrate the effect of the scan size on the multifractal characteristics of the films. The multifractal spectra (singularity strength), generalized multifractal dimension, and mass exponent functions were used as the parameters for multifractal characterizations and the following major findings were reported.

i. The profiles of multifractal spectrums, f(α), were seen to significantly depend on the scan size of the AFM imaging. It was seen that at small scan sizes the spectra are skewed to the right with a left hook whereas the spectrum at large scan size (10 × 10 μm) exhibits a nearly semi-oval shape. These results imply that AFM images obtained at small scan sizes exhibit less multifractal behavior as compared to those obtained at large scan sizes.

ii. The generalized multifractal dimension (D(q)) as a function of the moment order (q) revealed that images obtained at a scan size of 3 × 3 μm revealed a well-defined profile with the three fractal dimensions observable. On the contrary, for images obtained at larger scan sizes, the regions were not clear. The ordering of the generalized fractal dimension further showed that images obtained at 5 × 5 μm scan sizes exhibited nearly constant values for the fractal dimension, which indicates that the multifractal behavior at this scan size was the least. It was therefore concluded that AFM imaging should be taken at the scan sizes at which multifractality is less for accurate and reliable results.

iii. The mass exponent functions for the three scan sizes showed that at large scan sizes of the 10 × 10 μm, the profiles exhibit the clearest region of inflection unlike for the images taken at smaller scan sizes whose inflection regions were not very clear.

iv. From these results, it can be deduced that AFM images obtained at large scan sizes capture larger surface structures and hence higher roughness characteristics. However, such images are of lower quality and compromise on the quality of the topography details of the image [27]. It is therefore important to choose the correct AFM imaging parameters for thin films since they affect the fractal behavior of the images.

To investigate the influence thickness of gold thin films grown via electron beam evaporation, Yadav et al. [28] performed an MFDFA on AFM micrographs of the thin films. In the MFDFA method, the fluctuation usually expressed as $F_q(n)$ versus the window scale n for various values of the moment order, q was used to illustrate the influence of the films' thickness on the fractal evolution (see Chapter 3). The plot showed that (log $F_q(n)$ vs. log n) for every value of q, there exists a power law scaling between the fluctuation function and the window scale. The regression analysis using the least square method on the plot of log $F_q(n)$ versus log n gives the scaling exponent, generalized Hurst exponent function, h(q) against q and for the thicknesses investigated (20 nm, 50 nm, and 200 nm), the scaling exponent was decreasing with q. Additionally, the mass exponent $\tau(q)$ against q was also shown to reveal a nonlinear relationship, indicating that the gold thin films exhibited multifractality. However, the following points should be noted when using MFDFA method for multifractal characterizations.

i. For very large values of moment order, q, there is a small size of data and therefore the computation of the detrended fluctuation function, $F_q(n)$ becomes statistically less significant.

ii. As the absolute values of q become larger than 5, the scaling power law behavior of the surfaces becomes less clear, which means that beyond 5, the variation in scale gets destroyed.

iii. The scaling exponent h(q) against q is one of the important parameters in detecting structural differences in films grown at different conditions. For instance, for the gold films of different thicknesses reported by Yadav et al. [28], it was shown that there was a similar nonlinear and decreasing relationship between h(q) and q (for all the thicknesses) except for 200 nm thickness which deviated from the said relationship. It was discussed that the deviation is a result of the high roughness exhibited by the 200 nm thickness films.

Similar to the Al thin films analysis, gold thin films were also characterized using multifractal spectra ($f(\alpha)$), and the necessary parameters such as those shown in Tables 6.1 and 6.2 were computed. It was shown that all the spectra (of different thicknesses) were continuous functions of α. The spectra was also seen to strongly depend on the thickness of the gold thin films as per the study [28]. As far as the thickness of films and multifractal spectra are concerned, the following keynotes should be remembered.

i. The shape of the $f(\alpha)$ is strongly correlated with the thickness of the film thickness. For this study [28], Yadav et al. discussed that films with 20 nm and 200 nm exhibited spectra skewed to the right with a left-hook whereas those with 50 nm thickness were right-hooked.

ii. The increase in film thickness results into a wider multifractal spectrum. As such, wider spectra are indicators of the surface with high roughness, complex, and exhibiting singular characteristics.

iii. The values of minimum and maximum singularities (α_{min} and α_{max}) are also important in describing the influence of film thicknesses on multifractality. For instance, it has been shown that the minimum singularity decreases as the film thickness increases whereas the maximum singularity increases with the thickness of the films [13].

iv. The multifractal spectra width (earlier denoted as $\Delta\alpha$) is a measure of the range of height distribution of probabilities. As earlier stated, the smaller the value of $\Delta\alpha$, the weaker the multifractal behavior of the thin films, and the larger the $\Delta\alpha$, the stronger the multifractality of the surfaces. The influence of the thickness for pure metallic thin films has been described in detail by Khachatryan et al. [29] and shown that for Al films, the crystallinity, roughness, and complexity increases with the increase in the thickness. As such, the multifractal spectra width ($\Delta\alpha$) increasing with film thickness is an indication that films bearing larger thicknesses are stronger multifractals as compared to those with smaller thicknesses.

v. As per the mathematical formulation of multifractals, the minimum singularity (α_{min}) can be directly related with the maximum probability quantity (i.e. $P_{max} \approx \varepsilon^{\alpha min}$) whereas the maximum singularity (α_{max}) is related to the minimum probability measure as $P_{min} \approx \varepsilon^{\alpha max}$. In these formulations, ε is a very small quantity tending to zero [30]. It, therefore, means that the width of the spectra ($\Delta\alpha$) can indicate the probability range within which growth can occur.

vi. Based on the above descriptions for $\Delta\alpha$, then it means that $f(\alpha_{max})$ is the least quantity of growth probability whereas $f(\alpha_{min})$ is the maximum measure of growth probability. As such, Δf describes two important factors in thin films: (a) the height interval of the multifractal spectrum and (b) the ratio between the most and less populated areas as described in the literature [28].

6.3 MULTIFRACTAL STUDIES OF SPUTTERED METALLIC ALLOY THIN FILMS

In the preceding subsection, the multifractal descriptions of pure thin films have been presented. It is very clear to the readers that the deposition conditions, indeed, influence the multifractality of pure thin films. The thin film mechanisms occurring on the surface of the substrate can be detected via a multifractal characterization for thin films prepared via sputtering technology. Since these mechanisms vary with the composition of the substrate and target materials, then there is no doubt that metallic alloy thin films would behave differently from the pure metallic alloys [31]. In most cases, the metallic alloy films are used, especially when mechanical stability is necessary for their operations. The applications of pure metallic thin films are limited by mechanical strength and stability in various corrosive media. Metallic alloy thin films are therefore very necessary for advanced applications such as biomedical implants and wear resistant areas [32–34]. In this subsection, the authors describe some of the most important results on multifractal characterizations of metallic alloy thin films based on published literature.

In a study, published in *Chaos* journal in 2015, Yadav and his colleagues [35] investigated the multifractal behavior of nanostructured BaF_2 thin film surfaces. The thin films were grown on Si $\langle 1\ 1\ 1 \rangle$ wafer via electron beam evaporation method, which is very similar to the sputtering process, at room temperature and vacuum in the ranges of 10^{-6} torr. The BaF_2 were then irradiated in high energy ions of Ag in a high vacuum chamber. The obtained thin films were then imaged via AFM facility and then the images were characterized for mono- and multifractal properties using the respective algorithms as described in Chapter 3. The purpose of the analyses was to determine the influence of the ions' irradiation on the fractal characteristics of the alloy thin films. The MFDFA was utilized in studying the multifractal properties of BaF_2. The methodology uses a detrended fluctuation function $F_q(n)$ versus the scale size *n* plot on a double logarithmic scale beside the multifractal spectrum, scaling exponent (similar to generalized dimension), and mass exponent expressed as a function of the moment order.

The MFDFA analysis produces a double log $F_q(n)$ versus n plots, which in this case were shown to exhibit a power law relationship [35]. From these plots, the scaling/Hurst exponent function, similar to generalized fractal dimension, was determined and it exhibited a nonlinear decrease with the moment order, q for both irradiated and non-irradiated alloy films. The mass exponent for the films was also reported as a function of the moment order, q. It was shown that for all the films, the function exhibited a nonlinear relationship with q. The nonlinear function of the mass exponent is an indication of the multifractal nature of the height fluctuations of the films under investigation. In terms of multifractal spectra, the singularity strengths and widths of the spectra were determined and it was shown that α is higher for smooth surfaces, although, it is not a measure of roughness, it shows the variation of roughness with the length scale for the alloy thin films. In a similar study, the influence of the film's thickness on the multifractal characteristics was reported by Yadav et al. [13]. Similar to the above descriptions, the MFDFA measurements showed that the $F_q(n)$ function, h(q), spectrum and mass exponent functions exhibited the multifractal behavior. It was reported that the increase in the thickness of the LiF thin films resulted in shifting of the peak values of $f(\alpha)$ to the right and that the width of the spectra increased with the thickness of the thin films.

In a related study, the multifractal analysis on Ta_2O_5 thin films deposited via the electron gun method was reported [36]. The influence of the deposition pressure and annealing on the films was investigated. Atomic force micrographs were used for the multifractal characterizations using multifractal spectra and generalized fractal dimension parameters. All the multifractal spectra were humped and hooked to the left, indicating the multifractality of the thin films. The plots exhibited different widths and maximum spectral (fractal dimensions) values, which means that the influence of the annealing and pressure is significant. The multifractal spectra width, $\Delta\alpha$, was shown to be larger on films exposed to the annealing process. This means that annealing enhances multifractality, which can be associated with the formation of complex and singularity surfaces. The Δf values obtained for the spectra are measures of inhomogeneity, and it was shown increasing the annealing temperature results in larger values of the Δf parameter. The generalized fractal dimension parameters as functions of the moment order revealed a nonlinear decreasing plot, further indicating that the Ta_2O_5 exhibited the multifractal behavior. The influence of working pressure on the fractal behavior was not very significant in the study although further investigations would be necessary to establish any possible correlations at constant deposition/annealing temperature. Other studies have used multifractal characterizations to evaluate the influence of the annealing conditions of thin films such as ZnO and similar correlations deduced [37].

A more interesting study on the multifractal-annealing condition relationship for ZnO thin films was presented by Sun, Fu, and Wu in 2002 in an article published in *Physica A: Statistical Mechanics and its Applications* [38]. In the study, 200 nm ZnO thin films were deposited on Si (100) substrate via reactive sputtering technique and at a substrate temperature of 300°C. After deposition, the thin films were annealed in air at two different annealing temperatures (850°C and 950°C) for 1 hour after which the samples were characterized using AFM. The AFM imaging was based on contact mode imaging, scan area of 3×3 μm^2, and each image consisted of 512×512 pixels. The AFM imaging was able to obtain the grain sizes and it was seen that the annealing temperature led to an increase in the grain sizes of the ZnO films within these temperature ranges of annealing. Multifractal spectra properties of the as-deposited and annealed ZnO thin films were computed and discussed in relation to the annealing conditions. The largest values of α_{min} were obtained at the maximum annealing temperature whereas the smallest value was observed at the lowest temperature. It was observed that the values of the spectra decreased on annealing up to 850°C and then increased to 950°C. Similar observations were made for the α_{max}, $f(\alpha_{max})$, and $f(\alpha_{min})$ values. The multifractal spectra functions were all humped and nearly symmetrical about the peak value of $f(\alpha)$. The multifractal spectra widths ($\Delta\alpha$) values increased gradually from as-deposited to the maximum temperature of annealing indicating that annealing enhanced the multifractality of the ZnO thin films. The Δf values were largest on the as-sputtered films, positive for as-sputtered and films annealed at 850°C whereas the smallest and negative values were obtained at the maximum annealing temperature. This observation indicates that surfaces of the unannealed ZnO thin films consisted of a larger number of the lowest valleys than the number of the highest peaks. It was also shown that the root means square roughness increased with the annealing temperature and the spectral widths of the multifractal functions. These results can be explained by the fact that annealing enhances the energy of diffusion and nucleation of Zn and O ions on the surfaces of the substrate. Increase in nucleation results in grain growth leading to a larger interface width and widening of multifractal spectra widths. In a similar study, Mn-doped ZnO were deposited on silicon wafers via the magnetron sputtering process and annealed at different temperatures [39]. The multifractal spectra widths were shown to increase with the annealing temperature indicating stronger multifractals on annealing. On the other hand, the unannealed samples had a positive and largest value of Δf, which indicate the presence of valleys and defects before annealing. On annealing, the Δf were seen to decrease to negative values indicating better surface coverage and reduction of surface properties. The study also showed that increased crystallinity results into stronger multifractality of the thin films of metallic alloys.

Recently, Mwema, the main contributor of this book, co-authored an article with Shakoury et al. (2020) on the multifractal analysis of Zinc Sulphide (ZnS) thin films grown on BK7 glass substrates using RF magnetron sputtering technique [36]. In the study, the influences of argon gas flow rate and working pressure of the sputtering chamber were investigated. The crystallinity and fractal behavior of the ZnS thin films were evaluated in respect to the sputtering conditions. It was discussed that the crystallinity of the deposited films increased with the increase in argon gas flow rate and working pressure of the sputtering system. In fact, at the lowest argon gas flow rate of 20 sccm and working pressure of 4.4 mTorr, the obtained thin films did not reveal any peaks on the XRD system and were amorphous. In this article, Mwema was involved in the fractal and multifractal characterization of the ZnS thin films due to his expertise in fractal analyses of Al thin films. For multifractal measurements, the parameters used were generalized Hurst exponents (H_q), mass exponents, and the multifractal singularity spectrum. The generalized Hurst exponent plots versus the moment order can be related to the generalized fractal dimension and the plots were shown to exhibit similar nonlinear decreasing behavior for all the four samples investigated. The following important notes can be derived from the study on the influence of working pressure and argon gas flowrate in a sputtering process for alloy thin films as suggested by this article.

i. The ZnS thin films deposited in high working pressures and argon flow rates were associated with the largest values of Hurst exponent whereas films grown at low working pressures and flow rates had the smallest values of the Hurst exponents at all values of the moment order.

ii. The mass exponents function for the ZnS thin films at all values of moment order were shown to exhibit an inverse relationship with the working pressure and argon gas flow rate. The lowest values of the mass exponent function were obtained on samples prepared at the highest argon gas flow rates and working pressures whereas the highest values were reported on the films sputtered at the lowest argon gas flow rates and working pressures of the sputtering chamber. The authors discussed that the behavior of the mass exponent function can be related to non-uniformity of the surface roughness of the ZnS thin films obtained on the RF magnetron sputtering process.

iii. The multifractal singularity spectrum was obtained at various singularity strengths for the ZnS thin films. All the profiles of the singularity spectra were non-linear and were skewed to the right with short hooks to the left. From the spectra, the spectral values (fractal dimensions, $f(\alpha)$), multifractal spectrum widths, and singularity exponents associated with maximum and minimum growth probabilities were computed. It was determined that increasing the working pressure of the sputtering chamber and argon gas flow rate results in an increase in the widths ($\Delta\alpha$), which means that such surfaces exhibit stronger multifractal characteristics. The singularity index associated with the maximum probability growth (α_{min}) was nearly constant although the highest value was obtained on ZnS films sputtered at a high flow rate and working pressure.

Similar observations were reported for α_{max} although these values increased with the increasing pressure and flow rates. It can be generally stated that the largest values of singularity exponent (α) were obtained on films grown at the conditions of the high flow rate of the argon gas and working pressures. The results imply that films deposited at the highest working chamber pressures and argon flow rates exhibit stronger multifractal behavior.

iv. There is a direct relationship between the multifractality and crystallinity behavior of thin films of alloy metals as demonstrated in the publication for ZnS [36]. It can be seen that films exhibiting higher crystallinity have stronger multifractal characteristics. The argument can be used to describe the relationship between working pressure/and argon flow rates. As it was explained in Chapter 1, the working pressure influences the rate of deposition of the target material onto the substrate and the mean free path of the target material. On the other hand, argon gas is used to create an inert environment for the creation and sustenance of the plasma inside the chamber. The plasma may be referred to as the 'heartbeat of the sputtering process' and all the parameters of the sputtering method should contribute directly or indirectly into the plasma formation, sustenance, or both. When there is an increase in argon gas flow rate and working pressure inside the deposition chamber, there is enhanced dislodgement and deposition of the target atoms onto the substrate, which may lead to the formation of complex surfaces. At some conditions of pressure and argon flow rates, there occurs obstruction of free transfer of the target atoms onto the substrate surface. Additionally, at very high flow rates of argon gas and high pressure, some argon atoms get impinged and embedded onto the surface of the substrates causing the formation of the complex morphology of the thin films. As such, the multifractality tendency at high flow rates of argon gas and working pressures of the sputtering chamber is associated with the formation of complex, crystalline, and high interface width of the thin films.

The relationship between the crystallinity of thin alloy films and multifractal characteristics was illustrated by a study by Hosseinpanahi et al. [40] in an article titled, 'Fractal Features of CdTe Thin Films Grown by RF Magnetron Sputtering' and published in the powerful journal of *Applied Surface Science* in 2015. The CdTE films were sputtered at room temperature and different deposition times of 5, 10, and 15 minutes. An XRD characterization of the films at different times revealed that the crystallinity of sputtered of CdTe films is a function of the sputtering time. It was observed that at all the deposition times, the films exhibited polycrystalline characteristics with preferential orientation along the plane (111). It was shown that the preferential orientation was intense for films deposited at the longest time. A multifractal analysis was applied onto the AFM micrographs of the films and various parameters used to illustrate the influence of the deposition time to the fractal behavior of thin films. The different plots revealed that all the films exhibited multifractal characteristics. Most importantly, the multifractal spectra functions were shown to be skewed to the right with a left hook. The spectra width, $\Delta\alpha$, was shown to increase with the sputtering time, which means that $\Delta\alpha$ increased with the crystallinity of the CdTe thin films. Similarly, the values of Δf increased with the sputtering time and were all positive. This means that films deposited for short time exhibited a greater number of valleys than the peaks and that longer exposer of the substrates to the target atoms inside the sputtering chamber enhances surface coverage and reduces the chances of creating pores and other defects on the substrates. However, it can be generally observed that all the Δf values were negative, which means that the films consisted of a larger number of lowest valleys than the highest peaks. The results can be attributed to the short times (maximum time used was 15 minutes) to undertake the sputtering process. As per the experience of the authors, sputtering for a long time like 1-hour results in strong crystallinity, higher thickness, and full substrate coverage and therefore fewer surface defects on the thin films. This is the reason for the negative values of the Δf for the Al thin film characterizations in Tables 6.1 and 6.2 in the previous subsection. Figure 6.15 illustrates the concept of time of sputtering and its relationship to the sputtering and thin film formation. As shown, time influences three main processes of thin film formation during the sputtering: (i) the time of generation of the atom material from the target, (ii) travel of dislodged target material to the substrate surface, and (iii) diffusion of the atoms onto the substrate surfaces. There should be sufficient time for the plasma to be generated and to travel to the surface of the target and to dislodge enough atoms from that target. Additionally, there should be enough time for the atoms to travel from the target to the substrate through the impedance of chamber pressures and argon gas particles. Finally, the atoms should have enough time to impinge and condense on the surface of the substrate leading to formation of thin film structures.

The multifractal characterization in relation to the thickness of metallic alloy films has also been reported. A study by Ţălu and his co-authors [41] investigated the fractal characteristics of Tin-doped In_2O_3 using AFM. The films were deposited via electron beam deposition technique on glass substrates and had different film thicknesses of 100, 150, and 250 nm. In the study, the multifractal singularity

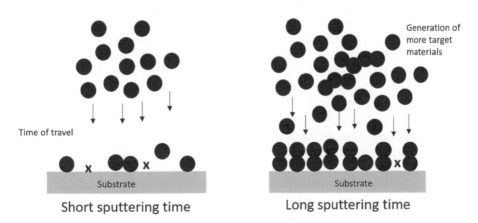

Short sputtering time   **Long sputtering time**

FIGURE 6.15 Illustrating the influence of the sputtering time on the sputtering process and formation of the thin films.

spectrum and generalized dimensions were used as the measure for multifractality. All the singularity spectra were humped and nearly symmetrical at different peak values of the spectrum. The results on the spectrum widths were quite interesting and contrary to what was described on the influence of film thickness for pure thin films of metallic materials. It was shown that as the thickness of the films increased, the multifractal width decreased considerably. This means the multifractality of the films was strongest on films with the lowest thickness. This result brings an interesting twist in the contextualization of the multifractal behavior of thin films. The interface width of the films was shown to decrease significantly with the film thickness from 100 nm (root mean square of 24.1 nm) to 250 nm (roughness of 4.94 nm). The values of the spectra (Δf) were shown not to exhibit a trend with the increasing film thickness. The generalized fractal dimensions, at $q = 0$, $q = 1$, and $q = 2$, were obtained and the following conclusions obtained: (i) the values at $q = 0$ were constant for all the film thickness and (ii) the largest value of fractal dimensions at $q = 1$ and $q = 2$ was obtained at a thickness of 150 nm. These results, therefore, imply the following as far as the multifractal behavior and film thickness applies for alloy materials.

i. The thickness does not directly influence the multifractality of thin alloy films, but rather, the growing evolution of the films. In some processes and conditions, the thickness of the films does not necessarily contribute to the complexity of the thin films but rather the driving conditions.

ii. The interface width or the root mean square roughness can be seen to play an important role on the multifractal behavior of the thin films of metallic alloy materials. As it can be shown in the plots (Figure 6.16), many studies can confirm that multifractal spectrum properties ($\Delta\alpha$ and Δf) exhibit a non-linear increasing relationship with interface width (root mean square roughness). These plots have been obtained from different studies for different thin films of alloy metals deposited via different techniques and process parameters. It further implies that there is a relationship between the lateral deviation and vertical roughness of surface features on thin films. It means that the lateral variation of the fractal dimensions across the surfaces of thin films is greatly influenced by the local quantity of the height features. However, this does not imply that the spectral parameters are direct measures of vertical roughness but indicate the variations, in terms of spectral length, of such features in the lateral direction of the surface planes of the thin films.

In another study, Ţălu, Morozov, and Yadav in 2019 [42] reported on the effect of reactive and argon gases on the DC magnetron sputtering of ITO thin films. The films were prepared at several combination of gases as follows: (i) argon gas, (ii) oxygen + argon, (iii) argon + oxygen + hydrogen, and (iv) argon + oxygen + hydrogen + nitrogen. The multifractal computations showed that the largest multifractal spectrum width was obtained on the films grown in presence of argon, oxygen, hydrogen, and nitrogen whereas the lowest value was obtained on films grown in presence of argon, oxygen, and hydrogen gases. The values of Δf were all positive and less than 1, which indicates that the surfaces had a greater number of valleys than peaks

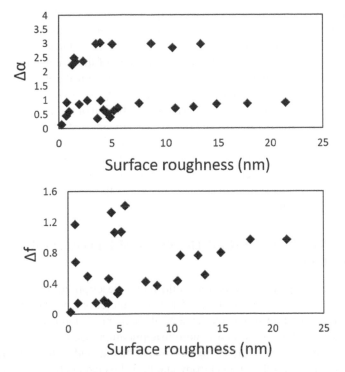

FIGURE 6.16 The relationship between the multifractal spectrum parameters and root mean square roughness values of various thin films. The data plotted on these graphs were obtained from the following references [5, 38, 43–52].

although not significant. The films with the lowest value of Δf was obtained when sputtering was undertaken at conditions of argon, oxygen, hydrogen, and nitrogen, which means that these surfaces were strongly multifractal. Interestingly, the surface roughness values in this study did not show a direct relationship as it has been reported in Figure 6.16. The finding can be related to the microstructure of the films obtained on the surface and the direct influence of the reactive gases onto the film formation. The films deposited in the presence of all the four gases were seen to exhibit the strongest multifractal character; which means that the formation of complex chemical structures has a stronger influence on the multifractality than the surface roughness. Therefore, in such cases, the multifractality can be described in terms of chemical compounds formed and their complexity. The influence of the reactive gases onto the structure of the ITO thin films was however not reported in the article and it is an interesting aspect to investigate in the future in relation to the multifractal characteristics of thin ITO films. It would also be interesting to evaluate the influence of the formation of complex chemical compounds during reactive magnetron sputtering on the statistical surface roughness of thin films for metallic alloy materials. The formation of the complex compounds during sputtering is the major difference between thin films for pure metallic and alloy metallic materials. It is obvious that the way the different target materials react or interact at the surface of the substrate will significantly affect the surface roughness and fractal character of the films. However, this obvious difference for pure and compound thin films is still not explored in the theory of sputtering of thin film deposition. For instance, during the formation of complex compounds of the thin film materials, some of the reactive target materials may result into the formation of localized precipitates on the surface of the substrates. Additionally, due to the nature of the sputtering process, non-uniform coverage of the substrate surfaces may lead to poor distribution of the compound products, especially in a multitarget sputtering process. This means that the surface of the substrate will be covered by significantly varying chemical compounds, which obviously may have a different texture and surface roughness, leading to varying surface roughness across the different lateral scales. It is therefore expected that deposition of compound alloy thin films will significantly affect the surface roughness and multifractal behavior of the thin film during sputtering.

6.4 MULTIFRACTAL DESCRIPTIONS OF SPUTTERED MULTILAYER THIN FILMS

From the preceding discussions, the multifractal descriptions for pure and alloy metallic thin films have been presented. There is an increasing need for the fabrication of multilayered thin film systems for enhanced mechanical stability and applications. The need for thin films for extreme condition applications such as in high temperature sensing, extreme wear conditions, high corrosive media, and so forth has pushed the thin film experts to develop high entropy and multilayered coatings. As discussed in the previous sections, multifractality is a subject of lateral complexity as well as the chemical complexity of the thin film coating. Preparation of multilayered thin films in a typical sputtering process involves the use of several target materials deposited in consecutive patterns to build layer-by-layer (LBL) films on the surface of a substrate. The process is very suitable for creating excellent films for optoelectronic display systems, biosensor, and drug screening components [53]. The LBL process of preparation of the films is likely to produce an interesting complexity in terms of roughness and fractal character.

A snippet search of the literature on multifractal descriptions of multilayer thin films reveals very limited information. In fact, there are no publications on multifractal studies of sputtered multilayer thin films. However, in one of the oldest studies by Sanchez et al. [54], a multifractal analysis was presented for multilayer GeAl thin multilayer films produced via laser irradiation process. The transmission electron microscopy was used to image the grown thin films and then the images were analyzed for multifractal properties using generalized fractal dimension and multifractal spectra. It was demonstrated that patterns of the multilayer that occur through diffusion and subsequent rapid solidification exhibited multifractal behaviors. In another study, Zhao and Liu (2011) investigated the fractal character of multilayer films of Au-Ta produced via ion-beam mixing [55]. The study revealed that the crystalline structures formed in the multilayer films were fractals.

To further understand the complexity and distribution of surface roughness in sputtered multilayer thin films, efforts should be concerted toward the multifractal characterization of these films.

6.5 SUMMARY

In this chapter, multifractal descriptions of sputtered thin films have been presented. The descriptions were based on the original data and published articles by the authors. Additionally, the literature on descriptions from other authors has been presented. The multifractal descriptions were presented for pure, alloy, and multilayer metallic thin films. The multifractal descriptions for thin films have been mostly described using the following parameters:

i. Multifractal spectrum: From the spectrum, the multifractal spectrum width and heights of the spectrum (spectral fractal dimensions) are used to determine the multifractal character of thin films. It agreed that the wider the spectrum, the stronger the multifractal behavior of the thin films. It is also discussed that the smaller the values of the height of the spectrum, the lesser the valleys than the peaks on the surfaces of the thin films. The shape of the spectrum determines the fractal behavior of the thin films – generally, spectrum with the hump, and (nearly) symmetrical about the maximum values of the spectral values are said to be multifractal.

ii. Generalized fractal dimension: The multifractal films are said to exhibit non-linear decreasing generalized multifractal dimension functions. From the function, three important parameters are described, namely, capacity dimension, information dimension, and correlation dimension. Generally, the capacity dimension should be the largest whereas the correlation dimension should be the smallest for multifractal thin films.

iii. The mass exponent and scaling exponent functions are also used in the multifractal characterization of thin films. It is generally agreed that for multifractal surfaces, these functions exhibit nonlinear decreasing behavior. The functions indicate the height fluctuations on the surface of the thin films.

Key findings on multifractal studies of thin films have been described in this chapter and some of the findings include the following:

i. Generally, interface widths of thin films exhibit a direct relationship with the width of the multifractal spectrum.

ii. The multifractality of thin films for pure metallic materials depend on the crystallinity of the films. The stronger the crystallinity, the stronger the multifractality of the films.

iii. For most alloy (metallic) thin films, the multifractality is strongly influenced by annealing after deposition. Annealing has been shown to increase and transform the span (or height) of the multifractal spectrum from positive to negative values. This indicates that annealing enhances the lateral evolution of the surface structures, which leads to the formation of stronger crystallinity and reduction of valleys (surface defects).

iv. The multifractality of thin films depends on the lateral complexity of the thin films and processes or parameters of sputtering which enhances such evolutions that are strongly correlated to the multifractality of the films.

The multifractal behavior differences among pure, alloy, and multilayer sputtered thin films are expected due to the nature of the formation of the films. The fact that sputtering of pure thin films, such as Al, requires a single high-purity metallic target, implies that the processes occurring at the surface of the substrate are simple and may result into the formation of a single chemical film. However, for alloy and multilayer thin films, there are more than one target used in the sputtering process that means that the thin film formation processes are complex and exhibit stronger multifractal characteristics. It is noted that there is scarce literature on the multifractal characterization of sputtered multilayer thin films and the subject is not fully understood.

REFERENCES

[1] C. Liu, X. L. Jiang, T. Liu, L. Zhao, W. X. Zhou, and W. K. Yuan, "Multifractal analysis of the fracture surfaces of foamed polypropylene/polyethylene blends," *Appl. Surf. Sci.*, vol. 255, no. 7, pp. 4239–4245, 2009.

[2] Y. Xu, C. Qian, L. Pan, B. Wang, and C. Lou, "Comparing monofractal and multifractal analysis of corrosion damage evolution in reinforcing bars," *PLoS One*, vol. 7, no. 1, pp. 1–8, 2012.

[3] N. M. Zaitoun and M. J. Aqel, "Survey on image segmentation techniques," *Procedia Comput. Sci.*, vol. 65, no. Iccmit, pp. 797–806, 2015.

[4] M. Li, L. Wang, S. Deng, and C. Zhou, "Color image segmentation using adaptive hierarchical-histogram thresholding," *PLoS One*, vol. 15, no. 1, p. e0226345, Jan. 2020.

[5] S. Hosseinabadi, F. Abrinaei, and M. Shirazi, "Statistical and fractal features of nanocrystalline AZO thin films," *Phys. A Stat. Mech. its Appl.*, vol. 481, pp. 11–22, 2017.

[6] F. M. Mwema, E. T. Akinlabi, and O. P. Oladijo, *"Exploring the effect of rf power in sputtering of aluminum thin films-a microstructure analysis,"* Proceedings of the International Conference on Industrial Engineering and Operations Management, Birmigham City, UK. 2019, pp. 745–750

[7] S. Srivastav, S. Dhillon, R. Kumar, and R. Kant, "Experimental validation of roughness power spectrum-based theory of anomalous cottrell response," *J. Phys. Chem. C*, vol. 117, no. 17, pp. 8594–8603, 2013.

[8] F. M. Mwema, E. T. Akinlabi, and O. P. Oladijo, "Micromorphology of sputtered aluminum thin films: A fractal analysis," *Mater. Today Proc.*, vol. 18, pp. 2430–2439, 2019.

[9] J. Schindelin et al., "Fiji: An open-source platform for biological-image analysis," *Nat. Methods*, vol. 9, no. 7, pp. 676–682, Jul. 2012.

[10] A. Karperien, "FracLac for ImageJ, version 2.5," 2014. [Online]. http://rsb.info.nih.gov/ij/plugins/fraclac/FLHelp/Introduction.htm.

[11] M. Fabrizii, F. Moinfar, H. F. Jelinek, A. Karperien, and H. Ahammer, "Fractal analysis of cervical intraepithelial neoplasia," *PLoS One*, vol. 9, no. 10, p. e108457, Oct. 2014.

[12] R. Zuo and J. Wang, "Fractal/multifractal modeling of geochemical data: A review," *J. Geochemical Explor.*, vol. 164, pp. 33–41, May 2016.

[13] R. P. Yadav, S. Dwivedi, A. K. Mittal, M. Kumar, and A. C. Pandey, "Fractal and multifractal analysis of LiF thin film surface," *Appl. Surf. Sci.*, vol. 261, pp. 547–553, 2012.

[14] J. A. Thornton, "The microstructure of sputter-deposited coatings," *J. Vac. Sci. Technol. A Vacuum, Surfaces, Film.*, vol. 4, no. 6, pp. 3059–3065, 1986.

[15] F. M. Mwema, E. T. Akinlabi, O. P. Oladijo, and J. D. Majumdar, "Effect of varying low substrate temperature on sputtered aluminium films," *Mater. Res. Express*, vol. 6, no. 5, p. 056404, Jan. 2019.

[16] F. M. Mwema, E. T. Akinlabi, and O. P. Oladijo, "Fractal analysis of hillocks: A case of RF sputtered aluminum thin films," *Appl. Surf. Sci.*, vol. 489, pp. 614–623, Sep. 2019.

[17] F. M. Mwema, E. T. Akinlabi, and O. P. Oladijo, "Micromorphology and nanomechanical characteristics of sputtered aluminum thin films," *Materwiss. Werksttech.*, vol. 51, no. 6, pp. 787–791, Jun. 2020.

[18] F. M. Mwema, E. T. Akinlabi, O. P. Oladijo, and J. D. Majumdar, "Effect of varying low substrate temperature on sputtered aluminium films," *Mater. Res. Express*, vol. 6, no. 5, p. 056404, Feb. 2019.

[19] F. M. Mwema, E. T. Akinlabi, and O. P. Oladijo, "Evolution of surface roughness and mechanical properties of Sputtered Aluminum thin films," *J. Phys. Conf. Ser.*, vol. 1378, no. 3, p. 032093, Dec. 2019.

[20] S. Blacher, F. Brouers, and G. Ananthakrishna, "Multifractal analysis, a method to investigate the morphology of materials," *Phys. A Stat. Mech. its Appl.*, vol. 185, no. 1–4, pp. 28–34, Jun. 1992.

[21] F. M. Mwema, E. T. Akinlabi, and O. P. Oladijo, "Sustainability issues in sputtering deposition technology," *Proc. Int. Conf. Ind. Eng. Oper. Manag.*, pp. 737–744, 2019.

[22] F. M. Mwema, E. T. Akinlabi, O. P. Oladijo, M. P. Nikolova, and E. H. Yankov, "Microstructure and mechanical characterization of aluminum thin films on steel substrates," *Mater. Today Proc.*, vol. 18, pp. 2415–2421, 2019.

[23] F. M. Mwema, E. T. Akinlabi, and O. P. Oladijo, "Fractal analysis of hillocks: A case of RF sputtered aluminum thin films," *Appl. Surf. Sci.*, vol. 489, no. May, pp. 614–623, Sep. 2019.

[24] F. M. Mwema, E. T. Akinlabi, O. P. Oladijo, and S. Krishna, "Microstructure and scratch analysis of aluminium thin films sputtered at varying RF power on stainless steel substrates," *Cogent Eng.*, vol. 7, no. 1, pp. 1–12, 2020.

[25] F. M. Mwema, E. T. Akinlabi, O. P. Oladijo, S. A. Akinlabi, and S. Hassan, "A multifractal study of Al thin films prepared by RF magnetron sputtering," in *Lecture Notes in Mechanical Engineering*, Mokhtar Awang, Seyed Sattar Emamian, and Farazila Yusof, Eds., 2020, pp. 687–694.

[26] F. M. Mwema, E. T. Akinlabi, O. P. Oladijo, S. A. Akinlabi, and S. Hassan, "Effect of AFM scan size on the scaling law of sputtered aluminium thin films," in *Advances in Manufacturing Engineering, Lecture Notes in Mechanical Engineerin*, Mokhtar Awang, Seyed Sattar Emamian, and Farazila Yusof, Eds., 2020, pp. 171–176.

[27] F. M. Mwema, E. T. Akinlabi, and O. P. Oladijo, "Dependence of fractal characteristics on the scan size of atomic force microscopy (AFM) phase imaging of aluminum thin films," *Mater. Today Proc.*, vol. 26, pp. 1540–1545, 2020.

[28] R. P. Yadav, U. B. Singh, A. K. Mittal, and S. Dwivedi, "Investigating the nanostructured gold thin films using the multifractal analysis," *Appl. Phys. A*, vol. 117, no. 4, pp. 2159–2166, Dec. 2014.

[29] H. Khachatryan, S.-N. Lee, K.-B. Kim, and M. Kim, "Deposition of Al thin film on steel substrate : The role of thickness on crystallization and grain growth," *Metals (Basel).*, vol. 9, no. 1, p. 12, 2019.

[30] W. Wang, A. Li, X. Zhang, and Y. Yin, "Multifractality analysis of crack images from indirect thermal drying of thin-film dewatered sludge," *Phys. A Stat. Mech. Its Appl.*, vol. 390, no. 14, pp. 2678–2685, 2011.

[31] F. M. Mwema, E. T. Akinlabi, O. P. Oladijo, and A. D. Baruwa, "Advances in powder-based technologies for production of high-performance sputtering targets," *Mater. Perform. Charact.*, vol. 9, no. 4, p. 20190160, Apr. 2020.

[32] C. H. Chang, C. B. Yang, C. C. Sung, and C. Y. Hsu, "Structure and tribological behavior of (AlCrNbSiTiV)N film deposited using direct current magnetron sputtering and high power impulse magnetron sputtering," *Thin Solid Films*, vol. 668, no. October, pp. 63–68, 2018.

[33] M. A. Tunes and V. M. Vishnyakov, "Microstructural origins of the high mechanical damage tolerance of NbTaMoW refractory high-entropy alloy thin films," *Mater. Des.*, vol. 170, p. 107692, May 2019.

[34] D. Raoufi, "Morphological characterization of ITO thin films surfaces," *Appl. Surf. Sci.*, vol. 255, no. 6, pp. 3682–3686, 2009.

[35] R. P. Yadav, M. Kumar, A. K. Mittal, and A. C. Pandey, "Fractal and multifractal characteristics of swift heavy ion induced self-affine nanostructured BaF 2 thin film surfaces," *Chaos An Interdiscip. J. Nonlinear Sci.*, vol. 25, no. 8, p. 083115, Aug. 2015.

[36] R. Shakoury et al., "Multifractal and optical bandgap characterization of Ta2O5 thin films deposited by electron gun method," *Opt. Quantum Electron.*, vol. 52, no. 2, p. 95, Feb. 2020.

[37] K. Ghosh and R. K. Pandey, "Annealing time induced roughening in ZnO thin films: A fractal and multifractal assessment," *Mater. Sci. Semicond. Process.*, vol. 106, no. October 2019, 2020.

[38] X. Sun, Z. Fu, and Z. Wu, "Multifractal analysis and scaling range of ZnO AFM images," *Phys. A Stat. Mech. its Appl.*, vol. 311, no. 3–4, pp. 327–338, Aug. 2002.

[39] C. Y. Ma, W. J. Wang, S. L. Li, C. Y. Miao, and Q. Y. Zhang, "Multifractal, structural, and optical properties of Mn-doped ZnO films," *Appl. Surf. Sci.*, vol. 261, pp. 231–236, Nov. 2012.

[40] F. Hosseinpanahi, D. Raoufi, K. Ranjbarghanei, B. Karimi, R. Babaei, and E. Hasani, "Fractal features of CdTe thin films grown by RF magnetron sputtering," *Appl. Surf. Sci.*, vol. 357, pp. 1843–1848, 2015.

[41] Ş. Ţălu, S. Stach, D. Raoufi, and F. Hosseinpanahi, "Film thickness effect on fractality of tin-doped In2O3 thin films," *Electron. Mater. Lett.*, vol. 11, no. 5, pp. 749–757, Sep. 2015.

[42] Ş. Ţălu, I. A. Morozov, and R. P. Yadav, "Multifractal analysis of sputtered indium tin oxide thin film surfaces," *Appl. Surf. Sci.*, vol. 65, no. 3, pp. 294–300, Apr. 2019.

[43] H. Fu, W. Wang, X. Chen, G. Pia, and J. Li, "Fractal and multifractal analysis of fracture surfaces caused by hydrogen embrittlement in high-Mn twinning/transformation-induced plasticity steels," *Appl. Surf. Sci.*, vol. 470, no. October 2018, pp. 870–881, Mar. 2019.

[44] F. Hosseinpanahi, D. Raoufi, K. Ranjbarghanei, B. Karimi, R. Babaei, and E. Hasani, "Fractal features of CdTe thin films grown by RF magnetron sputtering," *Appl. Surf. Sci.*, vol. 357, pp. 1843–1848, Dec. 2015.

[45] R. P. Yadav, S. Dwivedi, A. K. Mittal, M. Kumar, and A. C. Pandey, "Analyzing the LiF thin films deposited at different substrate temperatures using multifractal technique," *Thin Solid Films*, vol. 562, pp. 126–131, Jul. 2014.

[46] Ş. Ţălu, I. A. Morozov, and R. P. Yadav, "Multifractal analysis of sputtered indium tin oxide thin film surfaces," *Appl. Surf. Sci.*, vol. 484, no. April, pp. 892–898, Aug. 2019.

[47] A. Modabberasl, M. Sharifi, F. Shahbazi, and P. Kameli, "Multifractal analysis of DLC thin films deposited by pulsed laser deposition," *Appl. Surf. Sci.*, vol. 479, no. February, pp. 639–645, Jun. 2019.

[48] G. Liu, L. Wu, and F. Zhang, "Multifractal spectra of atomic force microscope images of lanthanum oxide thin films deposited by electron beam evaporation," *Mater. Sci. Semicond. Process.*, vol. 31, pp. 14–18, Mar. 2015.

[49] Ş. Ţ lu, Z. Marković, S. Stach, B. Todorović Marković, and M. Ţ lu, "Multifractal characterization of single wall carbon nanotube thin films surface upon exposure to optical parametric oscillator laser irradiation," *Appl. Surf. Sci.*, vol. 289, pp. 97–106, 2014.

[50] D. Raoufi, H. R. Fallah, A. Kiasatpour, and A. S. H. Rozatian, "Multifractal analysis of ITO thin films prepared by electron beam deposition method," *Appl. Surf. Sci.*, vol. 254, no. 7, pp. 2168–2173, 2008.

[51] Ş. Ţălu et al., "Multifractal spectra of atomic force microscope images of Cu/Fe nanoparticles based films thickness," *J. Electroanal. Chem.*, vol. 749, pp. 31–41, Jul. 2015.

[52] M. Nasehnejad, M. Gholipour Shahraki, and G. Nabiyouni, "Atomic force microscopy study, kinetic roughening and multifractal analysis of electrodeposited silver films," *Appl. Surf. Sci.*, vol. 389, pp. 735–741, Dec. 2016.

[53] I. Lee, "Patterned multilayer systems and directed self-assembly of functional nano-bio materials," in *Multilayer Thin Films*, Gero Decher and Joseph B. Schlenoff, Eds., vol. 2, Weinheim, Germany: Wiley-VCH Verlag GmbH & Co. KGaA, 2012, pp. 985–1001.

[54] A. Sanchez, R. Serna, F. Catalina, and C. N. Afonso, "Multifractal patterns formed by laser irradiation in GeA1 thin multilayer films," *Phys. Rev. B - Condens. Matter Mater. Phys.*, vol. 46, no. 1, pp. 487–490, 1992.

[55] M. Zhao and B. X. Liu, "Fractal patterns in Au-Ta multilayer films upon ion beam mixing," *Philos. Mag. Lett.*, vol. 91, no. 3, pp. 229–236, 2011.

7 Fractal Prediction of Film Growth and Properties

7.1 INTRODUCTION

As discussed, and illustrated in the previous chapters, the fractal theory is a powerful tool in analyzing the scaling behavior of thin film surfaces. However, based on the existing literature so far, the interpretation of the mathematical results of the fractal analysis as it relates to the properties of thin films and surfaces is not fully understood and harmonized. This is because, despite the extensive publications on fractal theory and thin film scaling, there are existing controversies in terms of the interpretation of the fractal results. These controversies are evident on how different researchers discuss their fractal analyses on thin film surfaces due to their differences in the processing. For example, it is not clear so far what the Minkowski functionals indicate on the evolution of surface features of thin films during deposition. Researchers have reported contradicting implications of k-correlation or power law models during the interpretation of PSD functions of thin film surfaces. Another example of the controversy is on the multifractal analyses of thin films; as it was demonstrated, the formulations are complex with so many parameters and are not straightforward. It is, therefore, conspicuous in the literature that most researchers avoid the use of multifractal analysis in thin film descriptions.

It was noted that most of the fractal techniques are step-by-step methods and can be easily implemented on a computer program. Being a repeatable and reproducible procedure, then it means that the fractal methods can be used to predict the characteristics of the intended thin films. It is noted that there are no literatures employing fractal algorithms to predict the properties of thin films, which can be very useful in depositing the desired quality of the films. It is of no doubt that such predictions would minimize errors and enhance the quality of the deposited thin films. Hence, there is a need to develop reference fractal characterizations for thin film surfaces to assist in the interpretation of fractal studies and reduce the conflicting conclusions on different aspects of fractal characterizations. Additionally, such a resource would be useful in providing clear guidance on the relationship among various fractal parameters and the thin film deposition process. It has been reported, as discussed in the previous chapters, that fractal parameter describes the structural evolution on the surfaces of thin films during deposition, grain growth, and film formation. Due to the complex nature of the interdependence among the deposition (process) parameters, material properties, and growth mechanisms, it has been difficult to conclusively understand the relationship between lateral roughness and vertical roughness. As far as the relationship between these two roughness parameters are concerned, existing literature has given contradicting conclusions. However, the controversy has been attributed (in Chapter 5) to the complex influence and interactions of the process parameters.

The present chapter is based on the following three questions.

    i. Is it possible to categorize thin film surfaces based on surface morphology?

    ii. Is it possible to develop reference data for fractal analyses of thin film surfaces?

    iii. Is it possible to predict thin film formation using fractal theory?

These questions form the methodology/sections of this chapter and attempts are made to classify thin film surfaces according to the morphological observations and behavior, undertake dedicated fractal characterizations and statistical analyses on the different categories, and finally use the generated information to predict the evolution and growth of surface structures for thin films.

7.2 MORPHOLOGICAL CLASSIFICATION OF THIN FILM SURFACES

During the condensation of the thin film materials onto the surface of the solid substrate, there is a gradual modification of the properties (surface) of the substrate. The property of the resulting film is dependent on several factors, including (i) surface quality of the substrate, (ii) temperature of the substrate, (iii) the physiochemical properties of the substrate and the source/target material, (iv) method of deposition, and so forth. Based on these factors, there are various growth modes for the formation of thin films, namely (i) vertical stack growth, (ii) layer growth, (iii) island growth, and (iv) layer plus island growth. These modes were discussed in an early study by Cerofolini in 1978 [1] and summarized here since they are very critical in understanding the evolution and classification of morphologies of thin films.

    i. **Island growth mode,** also known as the Volmer–Weber mode of thin film growth, occurs when the atoms of the film material are bound to themselves stronger than the substrate materials. For example, polycrystalline metallic vapors being deposited on insulator substrates. As illustrated in Figure 7.1(a), the growth involves nucleation of single-crystal 3D islands, which are driven by the process conditions, expand to connect with each other. This is then followed by the coalescence of the islands leading to the formation of continuous thin films. The stages of this model (Figure 7.1(a) (i-iii)) are generally driven by the process factors affecting grain growth, surface morphology, texture, and stress states.

    ii. **Layer growth,** also known as the Frank–van der Merwe mode, is considered to be an ideal model for the growth of thin films. In this case, the atoms of the film materials are more attracted to the substrate than to themselves. It is also known as the layer-by-layer growth in which perfect lattice matching between the substrate and thin films being deposited is required for this model. Figure 7.1(b) shows the layer growth model for thin films and the mode occurs in auto-epitaxy and materials condensing more on refractory substrates. For example, layer-by-layer growth of lead sulfide (PbS) thin films through chemical bath solution synthesis was reported by Templeman et al. (2019) [2].

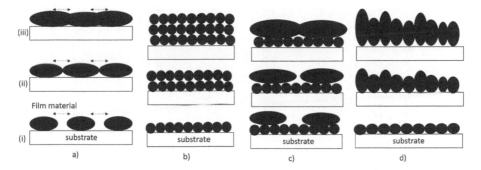

FIGURE 7.1 Growth modes in thin films: (a) island growth, (b) layer growth, (c) layer plus island, and (d) vertical stack growth.

iii. **Layer plus island growth** is another mode which is a composite of layer-by-layer and island modes and it is also known as the Stranski–Krastanov growth model. The mode represents an intermediate case between island and layer growth models. It occurs through, first, the formation of a three-dimensional layer on the substrate surface followed by the 3D islands. These islands then coalescence to form a continuous film on the layer film as shown in Figure 7.1(c).

iv. **Vertical stack growth** is shown in Figure 7.1(d) occurs when the atoms of the films are constrained on the surface of the substrate where they impinge and usually this mode occurs when there is no diffusion at the surface of the substrates. It is usually reported at very low substrate temperature. The atoms of the films tend to grow vertically on the substrate surface. This growth mode is usually associated with heterogenous formation of the surface structures.

Based on these modes, there are possible morphologies that can be obtained during thin film depositions. These morphologies include (i) columnar, (ii) ballistic, (iii) fibrous, and (iv) pile-up particles [3]. It should be noted that in most practical cases, a combination of these morphologies is likely to occur. As such, descriptions involving each one of these structures (morphologies) are mostly ideal and are important for numerical simulations/formulations.

7.2.1 Columnar Structure of Thin Films

Figure 7.2 shows a schematic illustration of the columnar structure of thin films. The film's structure appears like plates, which are perpendicular to the substrate surface across the cross section of the film–substrate interface.

These forms of morphologies are exhibited by different films as illustrated by various literature. For instance, Shin et al. (2006) [4] studied the growth mechanisms of ZnO thin films sputtered on p-type silicon substrates at varying sputtering substrate temperature. The grain sizes of the ZnO thin films were shown to be larger at the highest substrate temperature. The crystallinity and stability of the (0001)-oriented columnar structures were also shown to increase with increasing deposition

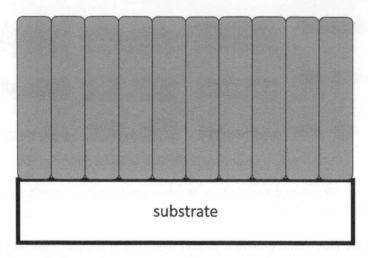

substrate

FIGURE 7.2 A schematic diagram of columnar structures of thin films.

temperature of the ZnO thin films. It was also shown that the upper region of ZnO columnar structures was more stable than the lower surface since at the lower region the grains of the films existed in a nonequilibrium state. The development of the columnar structures in the study was related to the island growth mechanism and the grains were convex in the upper region. In another older study, Karl H. Guenther (1982) described columnar and nodular growth of thin films in physical vapor deposition and the process parameters driving their formation and evolution [5]. The study idealized that the formation of the columnar structures in thin films is influenced by the melting point (and hence activation energy) of the target material. Based on the study [5] and references therein, it can be concluded that materials with high melting points such as Cr, Si, Ge, and compounds with high binding energy such as MgF_2 are likely to form columnar structured thin films. A low melting point implies high surface diffusion and structural relaxation, which hinders the build-up of columnar structures. Some of the materials not likely to form columnar structures include lead, tin, copper, and so forth. However, as reported for ZnO thin films, the surface energy and diffusion of atoms for driving the formation of columnar structures can be enhanced by increasing the substrate temperature during the film deposition process.

A study by Camach-Espinosa et al. [6] investigated the influence of sputtering time and substrate temperature on the morphology evolution of CdTe thin films. The CdTe films sputtered at room temperature exhibited columnar growth due to the equiaxed grains which were stack upon each other. The formation of the columnar structures at the low temperature in this study was attributed to the low mobility of atoms at low sputtering temperature. On contrary to the report by Guenther, increasing the substrate temperature results in less columnar structures of CdTe thin films. At higher temperatures, the columnar grains coalescence to form a nodular structure such that the film appears like a compact block at high substrate temperature (which was 250°C in this study [6]). Similar results for CdTe thin films were reported by Kulkarni et al. (2015) [7].

The columnar oriented structures have also been reported on silicon thin films prepared via radio-frequency thin films by Unagami, Lousa, and Messier (1997) [8]. The study investigated the influence of the argon gas pressure on the morphology of the silicon films. The Si films deposited at low argon gas pressure (10 mTorr) exhibited densely packed structures whereas the films deposited at a higher pressure of 30 mTorr exhibited a cauliflower-like morphology. The films deposited at 20 mTorr exhibited clear columnar structures with dome-shaped heads across the substrate–film interface. The Thornton SZM was used to discuss the evolution of the morphology of the films in the study. It was shown that columnar morphology obtained at 20 mTorr was related to zone 1 of the SZM. The formation of columnar structures in sputtered Si thin films was attributed to the oblique component of incident Si flux. As the argon gas pressure increases there is an increased extend of scattering of Si atoms, which enhances the oblique component of the Si flux. The concept of self-shadowing was also used to explain the formation of columnar structures in the study, in which target materials were stack on top of each other as described by the growth modes earlier. The readers are referred to Mauer and Vaßen article on the evolution of columnar structures in PVD thin films for further literature on the formation of columnar structures [9].

7.2.2 Ballistic Structure Models

In this model of thin film structure formation, the source (target) particles are assumed to travel in straight trajectories toward the surface of the substrate. On the substrate surface, they may collide with the already existing target atoms after which they relax to the next stable position or directly settle onto the substrate surface. In thermal evaporation, for example, the particles are assumed to move in straight lines at a certain angle of inclination. The deposition of the particles and relaxation depends on the geometrical positions of the target and substrate surface. In a sputtering process, the particles impinge the substrate surfaces randomly at a certain function of the distribution. This function is dependent on the geometry of target–substrate arrangement, the pressure of argon gas, temperature, and so forth.

When the particles hit the surface of the substrate, there occurs relaxation, surface diffusion, and desorption. These processes are detailed in literature [10,11] but it is important to note that they depend on the binding energy of the target materials. The ballistic deposition model is attributed to the formation of voids and porosity in thin films [12,13]. Figure 7.3 shows a model of the ballistic deposition of thin films. The target particle moves in a random trajectory to the surface of the already deposited atoms, where it migrated to a relaxed region.

7.2.3 Fibrous Thin Films

As the name suggests, the structure consists of fiber-like (elongated strings, whiskers, etc.) morphologies. According to SZMs, fibrous grains usually occur at zone T and the structure consists of whiskers. In this zone, there is negligible diffusion on the surface of the substrate and there are no voids and domes formed on the structures. For further understanding of the SZM, the readers are referred to the published

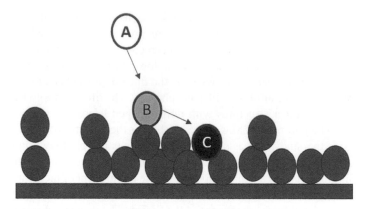

FIGURE 7.3   Ballistic model of thin film deposition.

literature [14,15]. Some of the films exhibiting these morphologies are chemically synthesized ZnO/AZO [16], polymeric sculptured films fabricated via PVD, and chemical vapor deposition (CVD) methods [17], sputtered MgO on SiO_2 substrates [18], etc. The concept of fibrous structures is extensively used in the creation of patterned thin film structures.

7.2.4 PILE-UP PARTICLES

This model predicts the formation of thin film materials during physical vapor deposition methods such as sputtering. In this case, atoms from the target material are randomly generated and deposited through piling on top of each other to create a continuous film. The model is very ideal but it represents a methodology to create and control thin film deposition through PVD methods.

7.3 FRACTAL CHARACTERIZATION OF THIN FILM MORPHOLOGIES

In this section, synthetic topographic images for the different possible morphologies and growth mechanisms were generated and their fractal characterizations were undertaken. The simulated surfaces are beneficial for theoretical analyses of artifact-free AFM images and are essential for the prediction of surface characteristics of specific structures [19]. The generation of the synthetic topography was conducted using Gwydion software (an open source software readily available on http://gwyddion.net/) and supported by the Department of nanometrology, Czech Metrology Institute [20,21]. The procedure for the generation of the synthetic topography micrographs are described in the manual of the software [21]. For each morphology category, a total of 10 images of varying scan sizes (up to 2000 μm square scan size) were generated. All the images were of the same resolutions of 512×512 pixels, height of 1.0 ± 0.00 pixels, coverage of 10.00, the inclination of $0.0°$, the direction of $0.0°$, a variance of 1.00, and at a weak relaxation mode [21].

7.3.1 COLUMNAR STRUCTURES

Figure 7.4 shows the artificial/synthetic atomic force microscopy (2D and 3D) of columnar structures of thin films. As shown, the columnar structures on AFM images appear like 'match sticks' with dome-shaped tops and faceted structures as it was demonstrated by Camacho-Espinosa et al. [6] and Mareus et al. [22]. The AFM images were then analyzed for fractal properties in the Gwydion software and the results are described next.

Figures 7.5 and 7.6 show the autocorrelation functions (A(r)) in vertical and radial orientations respectively of the AFM image in Figure 7.4. The autocorrelation function A(r) measures the interrelationship among the surface height features

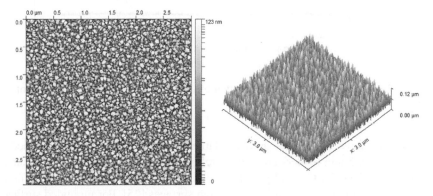

FIGURE 7.4 2D and 3D simulated AFM micrographs of columnar structures at a scan size of 3 μm × 3 μm.

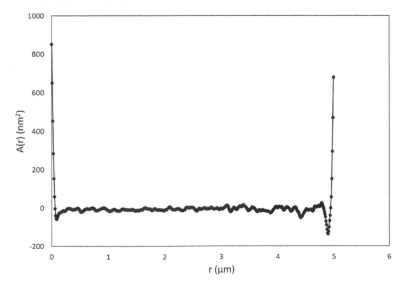

FIGURE 7.5 Areal autocorrelation function of the simulated columnar structures shown in Figure 7.4.

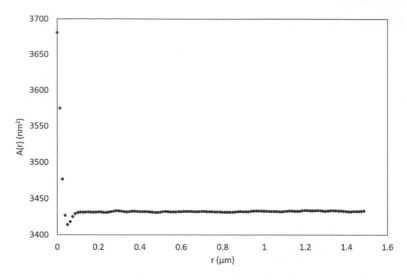

FIGURE 7.6   Radial autocorrelation function of the simulated columnar structures of thin films.

separated by a vector r. The autocorrelation function (A(r)) was described in Chapters 3 and 5 and readers are further referred to literature on its mathematical formulations [23–25]. Generally, A(r) exhibits exponentially decreasing behavior with the increasing r for self-affine and rough surfaces [25]. For mounded surfaces, A(r) also exhibits oscillations at the larger values of r. For perfectly columnar structures, as shown in Figure 7.5, the A(r) decreases exponentially to a dip and then increases sharply to maintain an oscillating behavior about a mean value of A(r). However, for radial A(r) in Figure 7.6, the function decreases and then remains nearly constant at the lowest value.

The height–height correlation function (also known as structure function) is also used to express the lateral roughness of the surface properties of thin films. Generally, the height–height function (H(r)) exhibits power law behavior at small values of r for self-affine surfaces. Additionally, at r = 0, the H(r) = 0 whereas for larger values of r, the power law scaling becomes dominant. For a perfectly columnar structure, the height–height function on a bi-logarithmic scale, shown in Figure 7.7, increases with increasing value of r and exhibits a power law behavior in this region beyond which the function nearly stagnates to exhibit a $2w^2$ relationship with r [26].

Another important method of determining the fractal characteristics of surfaces is the PSD function. The principles of PSD function were discussed in detail in Chapter 3 and extensively illustrated in literature [27–30]. The PSD as a function of the spatial frequency for the simulated columnar surface is shown in Figure 7.8 while the corresponding 2D fast Fourier (2D FFT) power spectrum is shown in Figure 7.9. The PSD function exhibits two major regions, the flat region (white noise region) at small spatial frequencies and the inverse power region at larger spatial frequencies. The 2D FFT spectrum does not show a clear distinction between the regions of high frequency and low frequency since the central region appears blurred. The observation

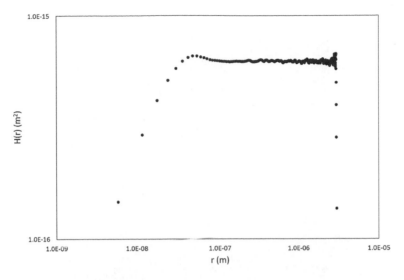

FIGURE 7.7 Height–height correlation function/structure of the columnar structures shown in Figure 7.3.

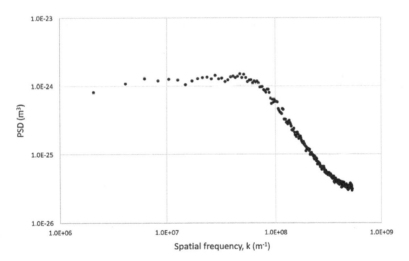

FIGURE 7.8 Power spectral density function of simulated columnar structures of thin films surfaces.

may be attributed to fully uniform columnar structures with 'dome-shaped' tops. The PSD plot does not exhibit any peak, which implies a self-affine surface and that the structure exhibits fractal characteristics.

The 2D Minskowski functionals are used to characterize the morphological features which cannot be studied through the classical methods of image analyses. Three functionals are usually used, namely, boundary length (S), Euler characteristic (X), and volume (V) (see Figure 7.10). These functions are based on the high and

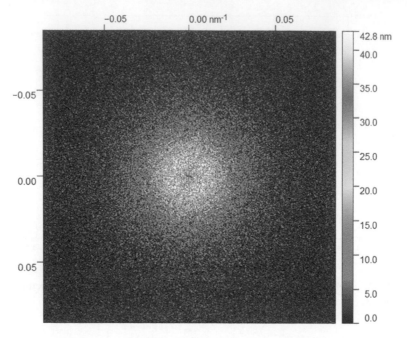

FIGURE 7.9   2D fast Fourier transform power spectrum of AFM image in Figure 7.3.

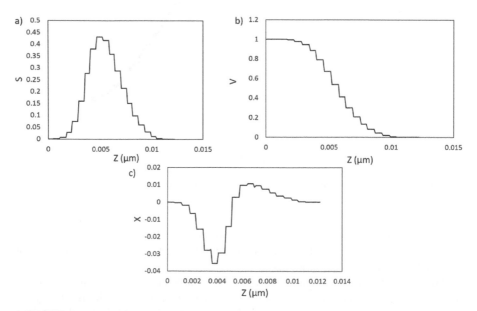

FIGURE 7.10   Minskowski functionals for the simulated columnar surfaces of thin films: (a) boundary length, (b) volume, and (c) connectivity.

low regions of the topographic (AFM) images. The Minskowski volume (V) measures the coverage across the surface of thin films. In this case, V is symmetrical about V = 0.5, which indicates dominance by self-affine columnar structures. The values of S are used to determine the global sizes of low and high domains of the structure and the nature of the surface morphology. The symmetrical profiles of boundary length (S) in this case (Figure 7.10) are characteristic of dominating columnar features on the surfaces. It can also be seen that the S functional in this case exhibit near Gaussian profile. The Minskowski connectivity (X) indicates the topological pattern of the AFM image. The X plot for the simulated columnar surface is dominated by the negative values of X and a small region of X in the positive range as shown in Figure 7.10.

A multifractal characterization of the simulated columnar surfaces is presented in Figure 7.11. The generalized multifractal dimension D(q) as a function of moment order (q) decreases gradually to the lowest value (D = 1.9), then increases and stagnates at about D = 1.95. It can be seen that in this case that the rule $D(0) > D(1) > D(2)$ cannot hold, which means that the structure does not exhibit multifractal characteristics. The mass exponent function ($\tau(q)$) against q is nearly a straight line; it does not clearly show the nonlinear relationship between the function and moment order. This observation indicates that the surfaces are not multifractal. Finally, the multifractal singularity spectrum, f(α), against the singularity exponent (α) exhibits a nonlinear decreasing spectrum with the singularity exponent. The spectral width ($\Delta\alpha$) in this case is zero as well as the Δf. It is also not possible to determine the value of f_{max} of the multifractal spectrum, which further confirms the lack of multifractality or

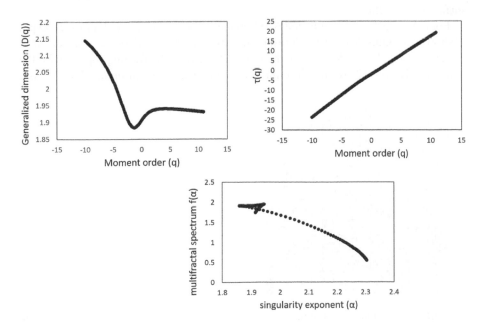

FIGURE 7.11 Multifractal properties of simulated columnar surfaces of thin films.

TABLE 7.1

The Fractal Dimensions of Simulated Surfaces of the Columnar Structure at different AFM Scan Sizes

Computation Method	Scan Size (µm)									
	3 × 3	5 × 5	20 × 20	50 × 50	100 × 100	500 × 500	800 × 800	1000 × 1000	1500 × 1500	2000 × 2000
Partition	2.834	2.835	2.832	2.831	2.831	2.832	2.834	2.831	2.835	2.832
Cube counting	2.624	2.624	2.624	2.612	2.622	2.621	2.628	2.607	2.627	2.634
Triangulation	2.706	2.706	2.711	2.685	2.703	2.712	2.715	2.696	2.702	2.704
Power spectrum	2.904	2.903	2.900	2.901	2.903	2.903	2.897	2.898	2.904	2.902

existence of weak multifractal behavior on the simulated columnar thin films. These observations are expected for a perfect columnar surface such as the one shown in Figure 7.4. Additionally, similar results are expected for AFM images dominated by columnar surfaces of thin films.

As defined in Chapter 2, the fractal dimension of fractal features is independent of the scale of magnification/measurement. To confirm the fractal nature of the simulated columnar surfaces, more artificial AFM images were generated at different scan sizes (5 µm × 5 µm, 20 µm × 20 µm, 50 µm × 50 µm, 100 µm × 100 µm, 500 µm × 500 µm, 800 µm × 800 µm, 1000 µm × 1000 µm, 1500 µm × 1500 µm, and 2000 µm × 2000). The summary of the fractal dimensions determined using the three common methods for the different scan sizes are shown in Table 7.1. It can be seen that the fractal dimension (D) is the same at all the scan sizes indicating that the surfaces are fractal.

For comparisons, the simulated AFM image at a scan size of 2000 µm × 2000 µm is shown in Figure 7.12. The corresponding fractal characteristics are presented in Figures 7.13 and 7.14. Apart from the scale differences, the fractal properties for the image in Figure 7.12 are similar to those of the image shown in Figure 7.4. This is a

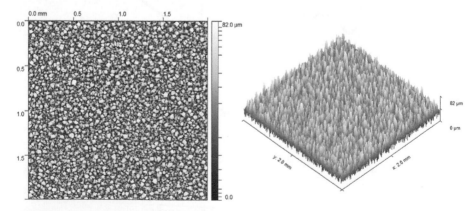

FIGURE 7.12 Simulated AFM image of columnar structures on the surface of the thin film at scan size of 2000 µm × 2000 µm.

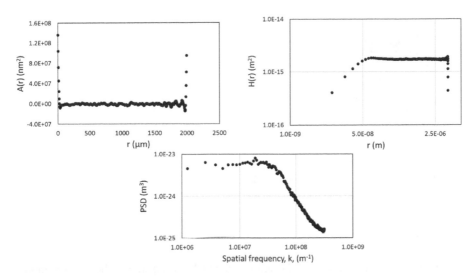

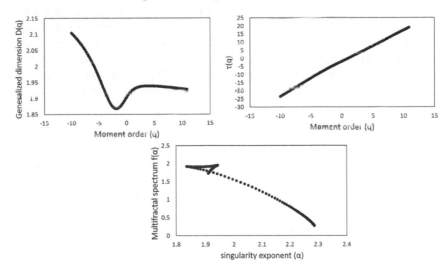

FIGURE 7.13 Autocorrelation, height–height correlation, and power spectral density functions for simulated AFM image shown in Figure 7.12.

FIGURE 7.14 Multifractal properties of the simulated columnar AFM image shown in Figure 7.12.

confirmation that the simulated columnar structures, in this case, exhibit self-affine characteristics, and the behavior is not affected by the scale of the imaging.

7.3.2 Ballistic Structures

Figure 7.15 shows a simulated AFM image of the ballistic model structure of thin film surfaces. The image shows a continuous island with 'ditches' to the adjacent islands. These ditches can be interpreted as the porosity/voids in ballistic thin film structures as reported in the literature of actual thin film surfaces [31].

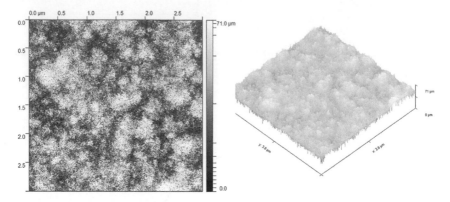

FIGURE 7.15 Simulated AFM micrograph of ballistic structures of thin film surfaces (scan size of 3 µm × 3 µm).

The fractal characteristics of the simulated AFM micrograph of the ballistic model of thin film surfaces are represented by autocorrelation (A(r)), height–height correlation (H(r)), and PSD functions shown in Figure 7.16. As shown, the A(r) function exhibits exponential decay at small values of r and then stable oscillations about a mean value of r. The H(r) function exhibits a power law correlation at small values of r and constant values at large r. Finally, the PSD function exhibits white noise behavior at small spatial frequencies (k), a transition zone, and an inverse power law at small values of k. However, the power law on the PSD profile is not steep as it may have been observed for columnar structures. These observations indicate the fractal nature of the ballistic model of thin film structures.

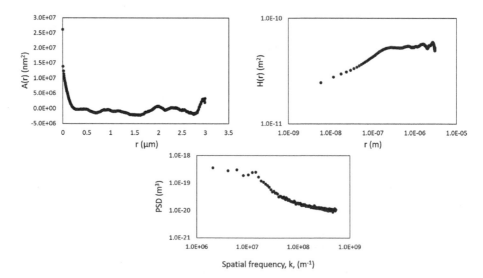

FIGURE 7.16 Autocorrelation (A(r)), height–height correlation (H(r))), and power spectral density (PSD) functions of the synthetic AFM image of the ballistic model of thin film surfaces.

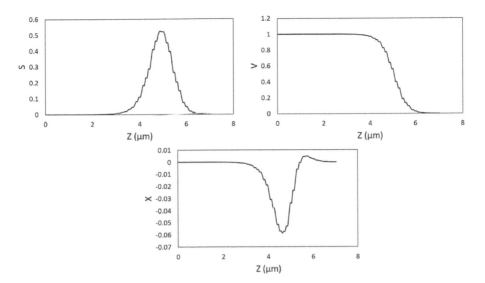

FIGURE 7.17 The Minskowski functionals (boundary length (S), volume (V), and Euler characteristic (X)) of the simulated AFM image of ballistic models of thin film surfaces.

The Minskowski functionals, boundary length (**S**), volume (**V**), and connectivity/Euler characteristics (**X**) are shown in Figure 7.17. As shown, the boundary length (S) is symmetrical about 5 µm, the Minskowski volume (V) is asymmetrical with a constant value of V = 1 of up to Z = 4 µm, whereas the connectivity (X) is dominated by negative values with a very small region of the plot in the positive region. It can also be seen that the boundary length functional (S) exhibits nearly a Gaussian profile.

Similar to Table 7.1, fractal dimensions were computed using the four techniques, partitioning, cube counting, triangulation, and power spectrum for simulated AFM models (ballistic) generated at the 10 different scan sizes. The average fractal dimensions across the different scan sizes were shown to vary insignificantly (standard deviation of about 0.009). This means that the fractal dimension was not affected by the scaling, which further confirms the fractality of the ballistic thin film surfaces.

The multifractal characteristics of the simulated ballistic structure of the thin film are shown in Figure 7.18. As seen, the multifractal generalized dimension D(q) function is decreasing gradually to q = 0, where a point of inflection occurs and there is a change in the slope of the function. It is also seen that D(q = 0) > D(q = 1) > D(q = 2), which indicates a multifractal behavior of thin films. The mass exponent (τ(q)) function exhibits nonlinear behavior and a point of inflection with the moment order, which is a characteristic of multifractality. The multifractal spectrum (f(α)) plots as a function of the singularity exponent (α) reveal a left-hooked shape. In this plot, parameters such as $\Delta\alpha$, Δf, and f_{max} can be estimated. These parameters are used as measures of multifractality in thin film surfaces, and therefore confirm the multifractality of the ballistic model of the films. The existence of multifractality is attributed to the complex nature of the ballistic structures – the presence of island structures with voids/porosity across the surface of the AFM image (Figure 7.15).

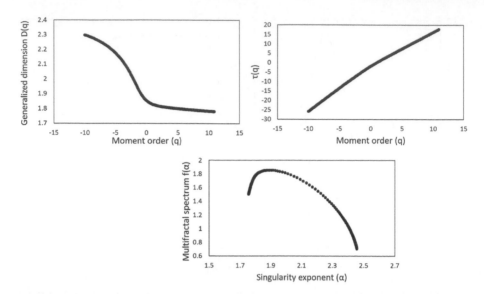

FIGURE 7.18 Multifractal properties the simulated AFM image of ballistic models of thin film structures.

7.3.3 FIBROUS MODEL OF THIN FILM STRUCTURES

Figure 7.19 shows a simulated surface of the fibrous structure of thin films. As seen in the 2D AFM topography, the surface consists of continuous and lateral string-like features. Such features are usually found in natural films such as collagen and proteins deposited on different substrate materials [32,33].

Figure 7.20 represents the autocorrelation function, height–height function, and PSD function for the simulated AFM image shown in Figure 7.19. The functions represent the fractal behavior of the surfaces. However, it is important to note that the height–height function does not exhibit normal oscillatory behavior at large

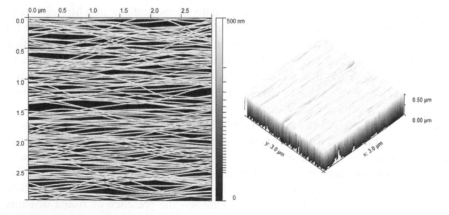

FIGURE 7.19 Simulated AFM fibrous surface model for thin films (scan size of 3 μm × 3 μm).

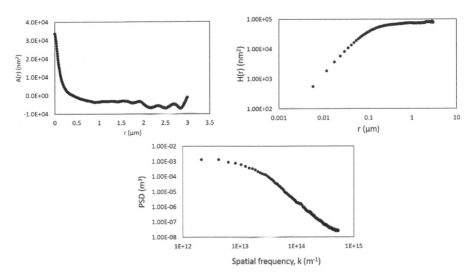

FIGURE 7.20 Autocorrelation (A(r)), height–height (H(r)), and power spectral density (PSD) function of the simulated AFM images of fibrous structures.

values of r. Additionally as seen on the PSD profile, the white noise region is very small and the transition region is not clear like it was observed for the previous two (columnar and ballistic) models. The fractal dimensions across different scan sizes of the fibrous AFM image exhibit insignificant change, indicating that the surfaces are fractal.

The Minskowski functionals for the simulated AFM image of fibrous thin film structures are shown in Figure 7.21. The boundary length (S) and Minskowski volume (V) are asymmetrical. The V exhibits a half-circle shape whereas the S functional is skewed to the right. The Euler characteristic does not reveal a clear transition between the lower and high domains. The multifractal characteristics of the fibrous surfaces are shown in Figure 7.22. The generalized fractal function D(q) exhibits decreasing behavior with moment order (q) with a slight transition at q = 0. Although D(q = 0) > D(q = 1) > D(q = 2) is true from the generalized multifractal dimension function, D(q = 1)-D(q = 2) ≈ 0 applies. This observation indicates a weak multifractality of the structure. It can also be seen that the mass exponent function (τ(q)) function increases with moment order; however, the inflection point is not clear as it was in the previous cases. The multifractal spectrum (f(α)) versus the singularity exponent (α) exhibits a small hook to the left of the profile. These observations suggest weak multifractality of the simulated fibrous surfaces of thin films.

7.3.4 PILE-UP PARTICLES

Figure 7.23 shows a simulated AFM image of pile-up particles for thin film surface. The model assumes that during deposition, particles are layered on top of each other (piled-up) to create a thin film. As shown, the AFM image of such surfaces consists of particles layered on top of each other and the structure consists of long features

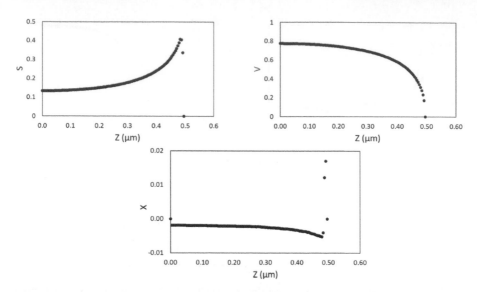

FIGURE 7.21 Minskowski functionals boundary length (S), Minskowski volume (V), and Minskowski connectivity (X) of the simulated AFM images of fibrous structures.

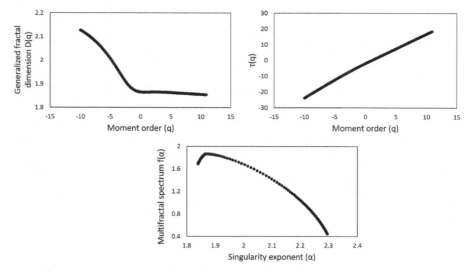

FIGURE 7.22 Multifractal characteristics of the simulated AFM image of fibrous thin film surfaces (generalized multifractal dimension D(q), mass exponent function (τ(q)), and multifractal spectrum f(α) functions).

with dome-shaped tops. Also, as shown in Figure 7.23, the surface has random voids across adjacent elongated pile-up structures. The fractal characterizations of the structure are presented in Figures 7.24 to 7.26. As shown in Figure 7.24, the autocorrelation function (A(r)) decays exponentially up to r = 0.4 μm, beyond which it oscillates at a mean value of about A(r) = −0.5 × 10³. The structure function (height–height

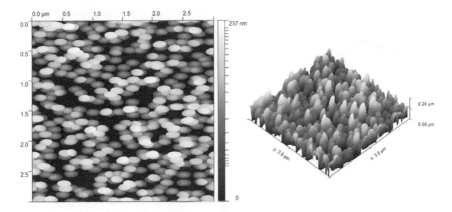

FIGURE 7.23 Representative simulated AFM image of pile-up particles for thin film depositions (scan size of 3 μm × 3 μm). [The topography (AFM) of pile-up particles shows long structures with dome-shaped tops. The surface is also characterized by voids within the adjacent pile-up structures.]

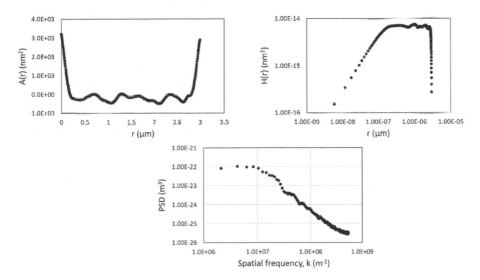

FIGURE 7.24 Autocorrelation (A(r)), height–height (H(r)), and power spectral density functions of pile-up particle structure of thin film deposition.

correlation function), H(r) exhibits power law behavior up to r = 1 × 10^{-7} μm, beyond which is stagnates with slight oscillations. The PSD function exhibits the white noise, transition, and highly correlated regions.

These results suggest that the pile-up structures are fractal. Additionally, fractal dimensions of the surfaces (shown in Figure 7.23) generated at different square scan areas (up to 2000 μm^2) revealed insignificant variation of fractal values with a scanning scale, implying the surfaces are fractal. The Minskowski functionals, boundary length (S), volume

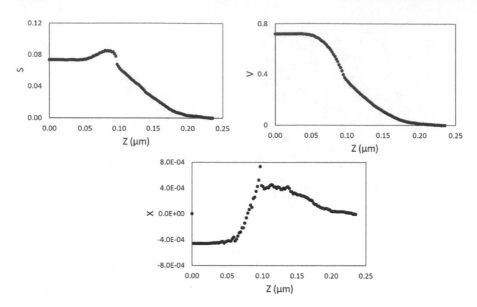

FIGURE 7.25 Minskowski functionals of the simulated AFM images of pile-up model for thin film deposition.

(V), and connectivity (X) are shown in Figure 7.25. It can be seen that S and V functionals are asymmetrical whereas the X is dominated by positive values due to the higher domain (resulting from piling of atoms on each other during the pile-up structure formation). The maximum values of S are skewed to the left and the V function exhibits an S-shape.

The multifractal characteristics of the pile-up structure are shown in Figure 7.26. The generalized multifractal dimension (D(q)) as a function of the moment order

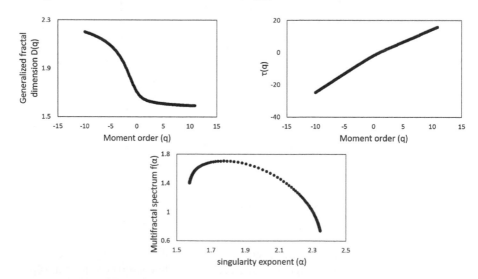

FIGURE 7.26 Multifractal characteristics of simulated AFM image of pile-up particle thin film deposition model.

decreases gradually up to q = 0 and then a change in gradient occurs. It can be seen that $D(q = 0) > D(q = 1) > D(q = 2)$ holds, which implies that the surfaces (Figure 7.23) are multifractal. The mass moment exponent $(\tau(q))$ versus q reveals a non-linearity relationship with an inflection point at q=0. This means that the structures exhibit strong multifractality. The multifractal spectrum, $f(\alpha)$, against the singularity exponent, α, reveals a humped profile skewed to the right with the f_{max} to the left of the profile. The parameters, $\Delta\alpha$, α_{min}, α_{max}, f_{min}, f_{max}, and Δf, which describe the multifractal properties of surfaces, can be easily determined. These results indicate that the structures of pile-up particles for thin film deposition are multifractal.

In the next section, methodology for predicting morphology and properties of the thin films based on the above fractal characterization is presented.

7.4 FRACTAL IMAGE ANALYSES, GROWTH, AND PROPERTY PREDICTION

The fractal characteristics described in Section 7.3 can be used to predict the behavior of deposited thin films. With the knowledge of fractal behavior of the ideal (synthetic) topographies such as those shown above, it is possible to classify the morphology of the 'unknown' structure. In a recent chapter by the authors of this book, a flowchart illustrating a methodology for morphology prediction through fractal knowledge was presented. The flowchart has been reused (with permission) in this section and it is illustrated in Figure 7.27.

In brief, the methodology involves two routes, namely (1) analyses of simulated AFM topography of thin film structures (columnar, ballistic, pile-up particles, and fibrous) and (2) validation route involving calibration, fractal analysis, and interpretation of results of actual (typical) thin film structure (in this case, CdTe thin films

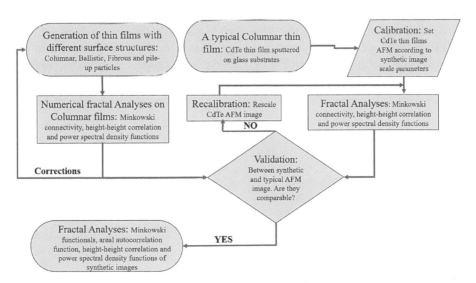

FIGURE 7.27 Fractal image analyses and growth prediction methodology (reused with permission from Springer Nature) [3].

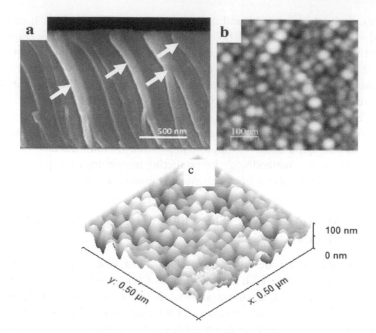

FIGURE 7.28 (a) Scanning electron microscopy image on the cross section of glass–CdTe thin films of the columnar structures (white arrows). (b) and (c) show the 2D and 3D of AFM image at the surface of the CdTe thin films and recalibrated image of CdTe thin films, respectively (reused with permission from Springer Nature) [3,6].

obtained via magnetron sputtering by reference [6]). The two routes should be taken for as many images as possible for statistical accuracy. Then, the fractal results of simulated surfaces are statistically compared to establish the relationship. For purposes of illustration, the CdTe thin films used in the methodology in Figure 7.27 are shown in Figure 7.28. The typical results of PSD comparing the simulated to the actual results are also shown in Figure 7.29. Statistical tools for error computation

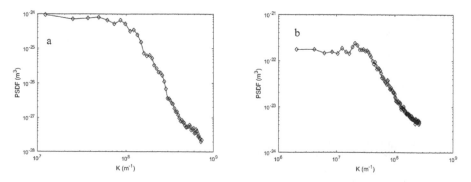

FIGURE 7.29 Power spectral density functions (PSDF) profiles on a log-log scale for (a) actual columnar structure of CdTe thin films as reported by reference [6] and (b) corresponding simulated columnar structure for thin film structures (reused with permission from Springer Nature) [3]

and the spread of data can be applied at this level to determine the accuracy of the methodology. It can be seen that the PSD profiles for simulations are correlated to the experimental profile in terms of behavior/trend and values.

For purposes of prediction of growth mechanisms and morphology of thin film structures, the methodology can be applied to actual thin film structures. This means that the latter route should be used and the results obtained can then be statistically compared to those obtained in Section 7.3.

7.5 SUMMARY

In this chapter, a classification of morphologies of thin films, with a bias to those mostly generated through PVD methods has been presented. Although the classification is not exhaustive, the four main models for describing the structures of PVD thin films are the columnar structure, ballistic structure model, fibrous structure, and pile-up particle structures. Other structure models, not discussed here, include diffusion, rods, and particle structure models for thin film growth. Then, based on the above classification, mono-fractal and multifractal characterizations of the structures have been undertaken and important observations are highlighted. All the models are shown to exhibit fractal characteristics, except the columnar and fibrous model that exhibit weak multifractal behavior (in fact ideal columnar structures were seen to exhibit mono-fractal behavior). A brief methodology for growth and property prediction of the morphology of thin films based on fractal characterization is also presented. The methodology has two routes, namely, the numerical (prediction) and experimental (validation) routes. Based on a typical illustration from an earlier work by the authors of this book, the methodology can be used to predict the growth and properties of the morphology of thin films from topographic information obtainable via scanning probe microscopy.

REFERENCES

[1] G. F. Cerofolini, "Morphology and morphological changes in thin films," *Thin Solid Films*, vol. 50, no. C, pp. 69–71, May 1978.

[2] T. Templeman et al., "Layer-by-layer growth in solution deposition of monocrystalline lead sulfide thin films on GaAs(111)," *Mater. Chem. Front.*, vol. 3, no. 8, pp. 1538–1544, 2019.

[3] F. M. Mwema, E. T. Akinlabi, and O. P. Oladijo, "Demystifying fractal analysis of thin films: A reference for thin film deposition processes," in *Lecture Notes in Mechanical Engineering: Trends in Mechanical and Biomedical Design*, Esther Titilayo Akinlabi, P. Ramkumar, and M. Selvaraj, Eds., 2021, pp. 213–222.

[4] J. W. Shin, J. Y. Lee, T. W. Kim, Y. S. No, W. J. Cho, and W. K. Choi, "Growth mechanisms of thin-film columnar structures in zinc oxide on p-type silicon substrates," *Appl. Phys. Lett.*, vol. 88, no. 9, p. 091911, Feb. 2006.

[5] K. H. Guenther, "*Columnar and nodular growth of thin films*," in *Proceedings SPIE 0346*, Arlington, USA. 1982, vol. 346, pp. 9–18.

[6] E. Camach-Espinosa, E. Rosendo, T. Díaz, A. Oliva, V. Rejon, and J. Peria, "Effects of temperature and deposition time on the RF- sputtered CdTe films preparation," *Superf. y vacío*, vol. 27, no. 1, pp. 15–19, 2014.

[7] R. R. Kulkarni et al., "*Properties of RF sputtered cadmium telluride (CdTe) thin films: Influence of deposition pressure*," in *AIP Conference Proceedings*, Jaipur, Rajasthan. 2016, vol. 1724, p. 020088.

[8] T. Unagami, A. Lousa, and R. Messier, "Silicon thin film with columnar structure formed by RF diode sputtering," *Jpn. J. Appl. Phys.*, vol. 36, no. Part 2, No. 6B, pp. L737–L739, Jun. 1997.

[9] G. Mauer and R. Vaßen, "Coatings with columnar microstructures for thermal barrier applications," *Adv. Eng. Mater.*, vol. 22, no. 6, p. 1900988, Jun. 2020.

[10] S. Müller-Pfeiffer, H.-J. Anklam, and W. Haubenreisser, "A generalized ballistic aggregation model for the simulation of thin film growth with special consideration of nodular growth," *Phys. Status Solidi*, vol. 160, no. 2, pp. 491–504, Aug. 1990.

[11] F. L. Forgerini and R. Marchiori, "A brief review of mathematical models of thin film growth and surfaces," *Biomatter*, vol. 4, no. 1, p. e28871, Jan. 2014.

[12] S. Models, "Ballistic aggregation models," in *Evolution of Thin Film Morphology*, Matthew Pelliccione and Toh-Ming Lu, Eds., vol. 108, New York, NY: Springer New York, 2008, pp. 121–141.

[13] V. A. Vasil'ev and P. S. Chernov, "Modeling the growth of thin-film surfaces," *Math. Model. Comput. Simulations*, vol. 4, no. 6, pp. 622–628, Nov. 2012.

[14] J. A. Thornton, "The microstructure of sputter-deposited coatings," *J. Vac. Sci. Technol. A Vacuum, Surfaces, Film.*, vol. 4, no. 6, pp. 3059–3065, 1986.

[15] I. Petrov, P. B. Barna, L. Hultman, and J. E. Greene, "Microstructural evolution during film growth," *J. Vac. Sci. Technol. A Vacuum, Surfaces, Film.*, vol. 21, no. 5, pp. S117–S128, Sep. 2003.

[16] C. Kumar et al., "Fibrous Al-Doped ZnO thin film ultraviolet photodetectors with improved responsivity and speed," *IEEE Photonics Technol. Lett.*, vol. 32, no. 6, pp. 337–340, Mar. 2020.

[17] S. Pursel, M. W. Horn, M. C. Demirel, and A. Lakhtakia, "Growth of sculptured polymer submicronwire assemblies by vapor deposition," *Polymer (Guildf).*, vol. 46, no. 23, pp. 9544–9548, Nov. 2005.

[18] L. Tang, Y. C. Feng, L. L. Lee, and D. E. Laughlin, "Electron diffraction patterns of fibrous and lamellar textured polycrystalline thin films. II. applications," *J. Appl. Crystallogr.*, vol. 29, no. 4, pp. 419–426, Aug. 1996.

[19] P. Klapetek and I. Ohlídal, "Theoretical analysis of the atomic force microscopy characterization of columnar thin films," *Ultramicroscopy*, vol. 94, no. 1, pp. 19–29, Jan. 2003.

[20] D. Nečas and P. Klapetek, "Gwyddion: An open-source software for SPM data analysis," *Cent. Eur. J. Phys.*, vol. 10, no. 1, pp. 181–188, 2012.

[21] P. Klapetek, D. Necas, and C. Anderson, "Gwyddion user guide," pp. 1–122, 2013.

[22] R. Mareus, C. Mastail, F. Anğay, N. Brunetière, and G. Abadias, "Study of columnar growth, texture development and wettability of reactively sputter-deposited TiN, ZrN and HfN thin films at glancing angle incidence," *Surf. Coatings Technol.*, vol. 399, no. June, p. 126130, Oct. 2020.

[23] S. Ţălu et al., "Micromorphology analysis of sputtered indium tin oxide fabricated with variable ambient combinations," *Mater. Lett.*, vol. 220, pp. 169–171, 2018.

[24] F. M. Mwema, E. T. Akinlabi, and O. P. Oladijo, "Fractal analysis of thin films surfaces: A brief overview," in *Advances in Material Sciences and Engineering. Lecture Notes in Mechanical Engineering.* Singapore: Springer, 2020, pp. 251–263.

[25] D. Nečas and P. Klapetek, "One-dimensional autocorrelation and power spectrum density functions of irregular regions," *Ultramicroscopy*, vol. 124, pp. 13–19, 2013.

[26] M. Pelliccione and T. M. Lu, *Evolution of Thin Film Morphology*, vol. 108. New York, NY: Springer New York, 2008.

[27] K. Ghosh and R. K. Pandey, "Power spectral density-based fractal analysis of annealing effect in low cost solution-processed Al-doped ZnO thin films," *Phys. Scr.*, vol. 94, no. 11, p. 115704, Nov. 2019.

[28] R. Gavrila, A. Dinescu, and D. Mardare, "A power spectral density study of thin films morphology based on AFM profiling," *Rom. J. Inf. Sci. Technol.*, vol. 10, no. 3, pp. 291–300, 2007.

[29] F. M. Mwema, E. T. Akinlabi, and O. P. Oladijo, "Dependence of fractal characteristics on the scan size of atomic force microscopy (AFM) phase imaging of aluminum thin films," *Mater. Today Proc.*, vol. 26, pp. 1540–1545, 2020.

[30] F. M. Mwema, E. T. Akinlabi, and O. P. Oladijo, "Effect of substrate type on the fractal characteristics of AFM images of sputtered aluminium thin films," *Mater. Sci.*, vol. 26, no. 1, pp. 49–57, Nov. 2019.

[31] J.-H. Kuang and H.-L. Chien, "The effect of film thickness on mechanical properties of TiN thin films," *Adv. Sci. Lett.*, vol. 4, no. 11, pp. 3570–3575, Nov. 2011.

[32] A. Stylianou and D. Yova, "Surface nanoscale imaging of collagen thin films by Atomic Force Microscopy," *Mater. Sci. Eng. C*, vol. 33, no. 5, pp. 2947–2957, Jul. 2013.

[33] L. Fang, C. Wischke, K. Kratz, and A. Lendlein, "Influence of film thickness on the crystalline morphology of a copolyesterurethane comprising crystallizable poly(ε-caprolactone) soft segments," *Clin. Hemorheol. Microcirc.*, vol. 60, no. 1, pp. 77–87, Jul. 2015.

Index